MW01032598

A Billion Butterflies

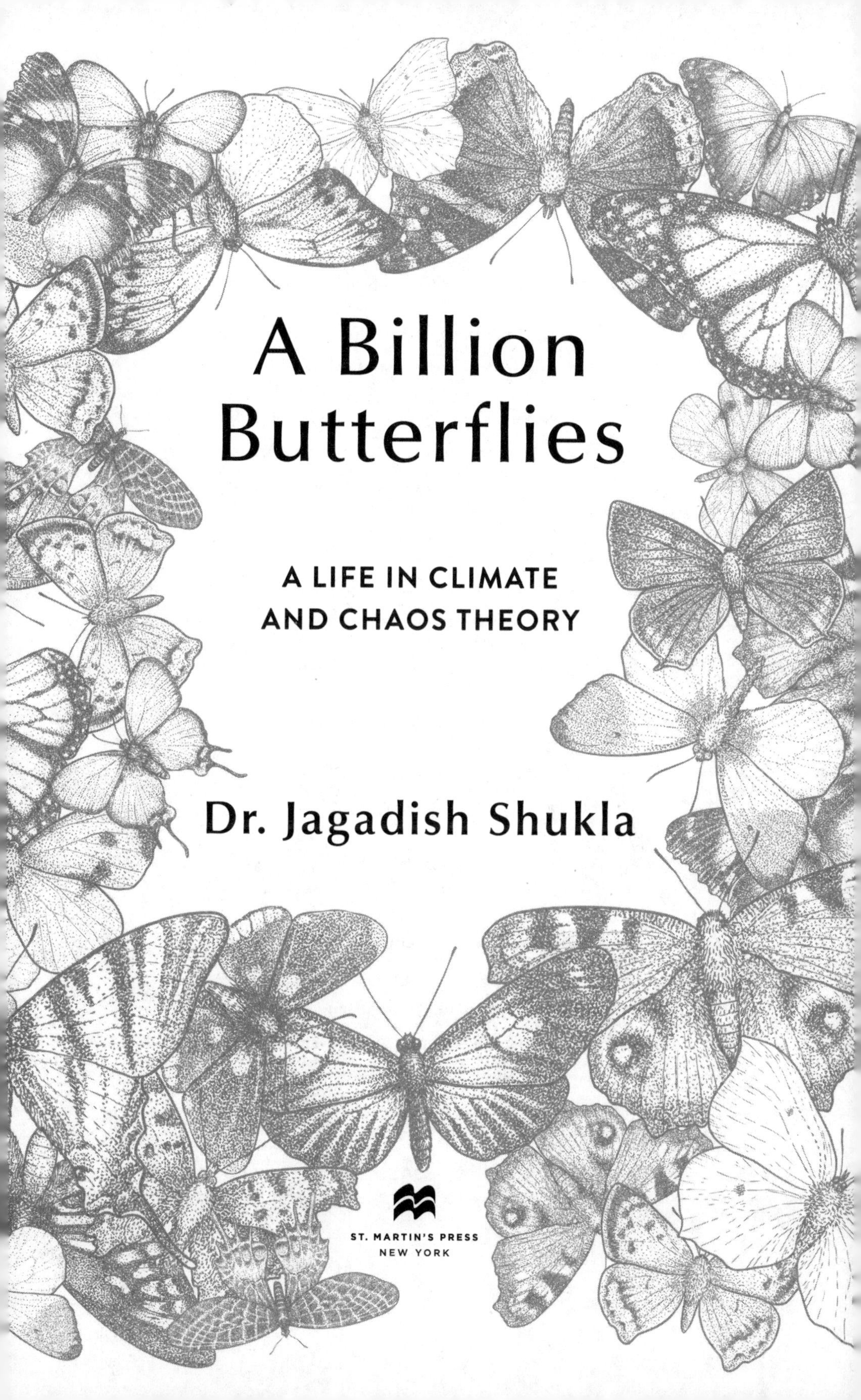

A Billion Butterflies

A LIFE IN CLIMATE AND CHAOS THEORY

Dr. Jagadish Shukla

ST. MARTIN'S PRESS
NEW YORK

First published in the United States by St. Martin's Press, an imprint of St. Martin's Publishing Group

This is a work of nonfiction in which dialogue has been reconstructed to the best of the author's recollection.

www.stmartins.com

Designed by Jen Edwards

Butterfly illustrations © Bodor Tivadar/Shutterstock

Insert photos provided courtesy of the author

The Library of Congress Cataloging-in-Publication Data is available upon request.

ISBN 978-1-250-28920-9 (hardcover)
ISBN 978-1-250-28921-6 (ebook)

First Edition: 2025

10 9 8 7 6 5 4 3 2 1

To my granddaughters Natasha, Aastha, and Aarushi,
in the hopes that they will adapt to and thrive in the
future climate they will inherit.

Contents

Prologue
Climate 101

At the large university where I work, I teach a class about global warming that is listed in the course catalog as Climate 101. Every year, I begin my very first lecture with the same deceptively simple request:

"Please raise your hand if you can tell me why it gets cold at night," I ask and watch as nearly every hand in the lecture hall goes up.

I point to an eager student, who can hardly wait to answer this very elementary question. "Because the sun has set, and the Earth is no longer receiving its heat"—or words to that effect—the student invariably tells me.

The other students nod in agreement. This is their answer too.

"Hmm," I say, pretending to consider how to evaluate this response. "For that incomplete answer, I will give you a B."

Suddenly, all those confident faces appear puzzled.

Perhaps you are puzzled too. Perhaps you had formulated the same response in your head. In fact, wherever I ask this question—at dinner parties, with family, or in conversation with new friends—nearly everyone gives me the same answer as an enthusiastic college freshman. They aren't wrong; rather, they are making a good guess based on an empirical observation. When the sun is visible in the sky, they feel warm. When the sun disappears, it's time to pull out the blankets.

But that's not the whole story, and the complete answer to this simple question can help us understand what climate is—and what is causing it to change.

As I tell my students, the sunset is certainly part of the answer. But to figure out why it is colder at night, we must remember that the Earth, like any other matter subject to the laws of physics, is *constantly* losing energy. Think of the way air in a room full of people grows warmer and warmer as all those human bodies radiate heat. The Earth does the same thing as those bodies. During the day, the sun provides a counterbalance to this energy loss, showering our planet with its warming rays. At night, the Earth continues to lose energy, and without the compensation of the sun, the temperature plummets.

On average, the Earth loses about 122,000 trillion watts of energy to space each year, which happens to be approximately equal to the amount of energy it receives from the sun. It is this balance—between outgoing and incoming energy—that determines the mean climate here on Earth. For nearly ten thousand years, the incoming and the outgoing energies were in such good balance that the global annual average temperature was a comfortable 14 degrees Celsius, making it possible for life to survive and humanity to thrive.

On Venus, this balance makes the annual average temperature 464 degrees Celsius. This isn't surprising, given that Venus is closer to the sun. But there is another big factor in addition to the energy a planet receives from the sun that influences climate: the chemical makeup of its atmosphere. On Venus, carbon dioxide accounts for 95 percent of the atmosphere. On Earth, our atmosphere is only about .04 percent carbon dioxide—or at least, it *was*.

In the past one hundred fifty years, humans chopped down many of Earth's carbon-sucking forests and began burning fossil fuels to heat their homes, power their factories, and propel their vehicles, releasing unprecedented amounts of CO_2 into the atmosphere. Like the glass walls of a greenhouse, CO_2 admits energy from the sun but prohibits energy from leaving the Earth. And so pretty quickly, our nicely balanced climate became imbalanced. In the past century, as the amount of CO_2 in the atmosphere has increased, Earth's global mean surface temperature has ticked up from 14 to 15 degrees Celsius.

This is called climate change. Climate change due to human activities is now firmly established by the observed facts and the laws of physics. The consequences of the phenomenon are becoming self-evident, but so are, I'd argue, the capabilities of the new generation of scientists to find a way forward.

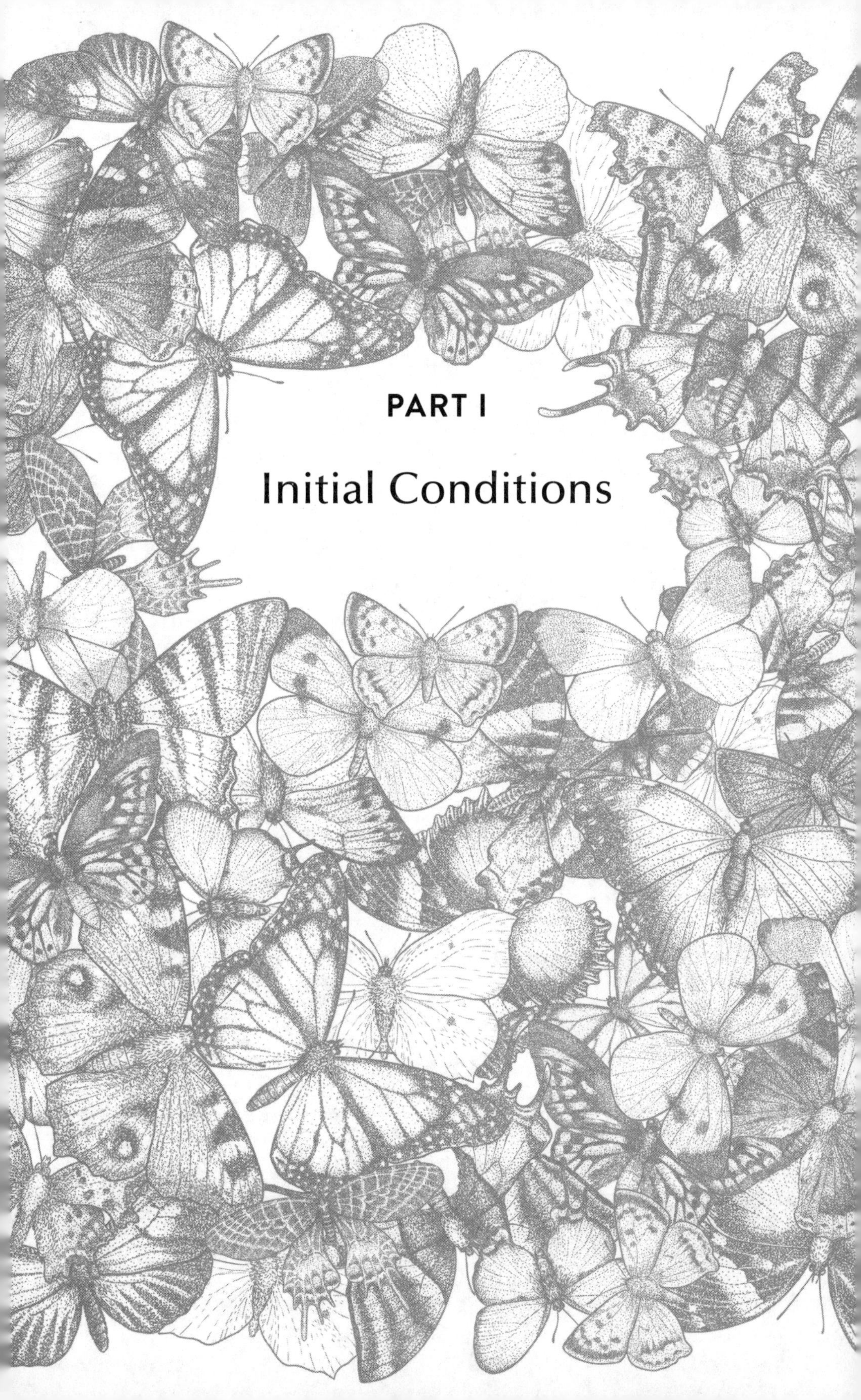

PART I

Initial Conditions

One

We were a few hundred feet above the Bay of Bengal when the windshield cracked. From my window, I looked out at the black water below, roiling with foamy whitecaps. Gunmetal clouds obscured any trace of a horizon, and bullets of rain pelted the thin plexiglass that separated us from the storm. We had been aboard this Electra turboprop plane—a "hurricane hunter," as it was known in the business—every day for a week, flying directly into the most severe monsoon storms we could track down, but this was the first moment I had felt fear.

I glanced over at Joachim Kuettner, a German atmospheric scientist and by far the most experienced member of our team. A veteran of field experiments around the world, Joach had volunteered to be the intermediary between the six scientists on board the plane and the pilot in the cockpit, relaying messages back and forth about our current coordinates or modifications to the flight path. It was Joach who had just calmly told us the windshield was cracked and that we would be dropping fuel and heading back to Calcutta (now called Kolkata). Now I tried to read the lines in his seventy-year-old face to ascertain how worried I should be, but his brow remained unfurrowed, his lips unpursed.

I scanned the rest of the team manning this flying laboratory; the world-renowned experts in the analysis and modeling of weather were perfectly silent. My Indian counterpart, a scientist named Dev Raj Sikka, studied his hands, which were folded tightly on the desk in front of him. Just behind Dev Raj, three scientists who had been chattering away all morning about the numbers on their screens had turned ashen and mute. Struck all at once by the same impulse, we began to pull mangoes and bananas from the baskets placed on board by the flight crew. With little else to do with the long moments before our demise, the six of us began peeling the fruit and taking large, joyless bites.

Just minutes before, we had all been focused on our various consoles, each executing his or her assigned task. As the chief scientist of the 1979 Summer Monsoon Experiment (MONEX for short), I kept track of the airplane's flight path, monitored the incoming data, and announced when it was time to launch another dropsonde, a small, cylindrical sensor attached to a parachute. As each of these instruments descended into the bay, it captured an enormous array of data—altitude, temperature, humidity, pressure, and wind speed—that it sent back via radio signal to the aircraft's recording devices as well as the computer monitors of the scientists on board. They pored over each batch of data as the plane heaved and stuttered through storm clouds that lined the sky like boulders in a creek.

The aircraft itself was also collecting information via nose and belly radar, multiple cameras, a radiometer, and a laser probe that penetrated the clouds. Meanwhile, our pilot maneuvered this flying laboratory in ever-descending elevations, carving the sky into a perfect layer cake of data until the plane was so close to the surface of the water, we could nearly count the fish below. It hadn't occurred to me until the windshield cracked that all these fancy instruments along with all my new friends could possibly wind up in that choppy water.

As I watched gasoline stream past my window, I thought about the village I had grown up in, only a half day's drive from the Bay of Bengal, where I had first experienced the power of a monsoon storm. I had never dreamed that one day I would be in an airplane with some of the most brilliant people in the world trying to understand what made one of these storms happen. Back then, I didn't even know *cars* existed, let alone airplanes, and the storms that passed through the village each year seemed as unknowable as the jackals that howled from the edge of the fields at night. They came or they didn't; no one tried to understand why.

During the weeks I had been in Calcutta for MONEX, communing with scientists from twenty-one other countries, many old friends and extended family members had made the journey from the village for a visit. It had been years since some of them had seen me; they remembered me as a barefoot kid kicking around a soccer ball made of old cloth and heaving cow dung into a bucket balanced on my head. Now I was wearing shiny shoes and a tie. I worked at NASA. My old neighbors looked as astonished by my new American life as I was. At thirty-five, I still felt very much like a boy from the village.

Maybe that's because so much of the work I did studying monsoons was the direct result of seeing how vital the rains were for the people in my rural village, farmers whose families went hungry in dry years and whose livelihoods were ruined by ones that were too wet. If humanity could get better at predicting monsoon precipitation, those farmers would not be so vulnerable; they could plan ahead for difficult times, plant earlier or later, sow different crops. But so much was still unknown about monsoons—what triggered the onset of the rains, what influenced the intensity of the storms—and MONEX was the largest effort ever undertaken to bridge the wide chasms in our knowledge.

In fact, the Electra was not the only airplane collecting data for MONEX that day. Two other aircraft, a NOAA P-3 and a NASA CV-990, were also flying at high altitudes, their own cameras, radars, and lasers interrogating the atmosphere. Sixteen ships were sailing the Indian Ocean taking readings of the sea's surface temperature.

Rocketsondes—sensors launched by rockets instead of being dropped from planes—were drawing arches through the stormy sky. Newly built micrometeorological towers on the east coast of India were recording wind turbulence. Weather balloons were drifting into the heavens at almost all hours of the day.

Ours was merely one phase of the Summer MONEX; earlier, there had been a two-month field experiment in the Arabian Sea. Scientists who had participated in the Winter MONEX in the South China Sea were already back at their home institutions attempting to wring meaning from the cache of data they had collected.

Incredibly, MONEX was just one of a half dozen regional experiments happening under the banner of 1979's Global Weather Experiment, an unprecedented effort to collect atmospheric and oceanic observations from around the world to improve the burgeoning field of numerical weather prediction. As ever-smarter supercomputers were giving humans the capacity to make millions of complicated calculations in mere seconds, we desperately needed a better understanding of the science behind those calculations—the interactions between air pressure, wind, mountains, humidity, radiation from the sun, cloud cover, the chemical composition of the atmosphere, and so on—not to mention the numbers to plug into them, vast tomes of data from places we had never collected it, such as the sky over the Bay of Bengal during the summer monsoon.

As the Electra shook with a sudden gust of wind, I gazed into the heart of the storm, hoping mightily that the plane, its passengers, and all this precious data would make it safely to the ground.

In 1961, a newly elected President Kennedy approached his science adviser, a man named Jerome Wiesner, and asked him for advice about his upcoming talk at the United Nations. Kennedy was looking for a flashy new project to propose, a scientific endeavor that could bring the world together at the height of Cold War tensions. Earlier in the year he had announced to Congress that the United States would send a man to the moon before the decade was over. Now he wanted to propose a similar challenge to his counterparts at the UN, something that would inspire cooperation and unity during a time of mistrust and deep division.

Wiesner, an electrical engineer who had made a name for himself developing microwave radar at MIT (he would later go on to serve as that institution's president), polled his colleagues at the university. What pressing scientific needs should mankind address next? he wanted to know. His friend Jule Charney had an idea—a global weather experiment. Every day for one year, people all over the world would take as many detailed observations of the weather as possible, using existing infrastructure like the weather stations humans had been building for centuries but also deploying new instruments to farther reaches than ever before.

At the time, Charney was in the midst of revolutionizing the field of meteorology. Instead of making weather predictions based on what had happened in the past, as scientists had done for years, Charney produced weather forecasts using only a supercomputer, the laws of physics, and initial conditions—that is, the atmospheric conditions at *this* very moment. The initial conditions of today are almost entirely responsible for the weather tomorrow and the day after that. But in 1961—before satellites began orbiting the globe and before many countries had invested in legitimate weather services—gathering those vital numbers was still a challenge. Charney envisioned this global weather experiment as a scientific marathon that would demonstrate that more observations and more accurate initial conditions enhanced supercomputer-derived weather forecasts. An effort like this would not only improve countless lives around the planet but also remind the world that despite cultural and political divides, humankind had more commonalities than differences. We all lived under the same sky, after all.

Wiesner knew a good idea when he heard one and on September 25, 1961, after calling for the creation of a UN peacekeeping force and a ban on tests of nuclear weapons, Kennedy proposed "further cooperative efforts between all nations in weather prediction and eventually in weather control."

In the years after Kennedy's speech, the UN's World Meteorological Organization began planning the Global Weather Experiment, which also became known as the Charney experiment.

The ambitious initiative took nearly two decades to organize. In that time, virtually every country on earth invested in more or better weather-observing technology. To fill the large gaps that existed in monitoring the oceans, scientists deployed weather buoys from boats and airplanes. Commercial ships were fitted with meteorological instruments, and forty research vessels were stationed in equatorial waters. Hundreds of upper-air stations readied radiosondes to send aloft. Two geostationary meteorological satellites were placed in fixed positions around the Earth, and several polar-orbiting satellites began circling the planet at a height of about eight hundred kilometers. Thousands of scientists and institutions were recruited to record, read, and analyze the data.[1]

By the beginning of 1979, the whole world was equipped to observe the actual weather from pole to pole, to watch in real time as the conditions of today turned into the conditions of tomorrow with fewer gaps and fewer guesses than ever before, forever enhancing our understanding of the complex systems that swirl around the globe and cause the weather that greets us every morning at our front doors.

In 1979, I was a young scientist working in monsoon research, and the Global Weather Experiment was a thrilling opportunity. I had just finished, about three years prior, a Ph.D. dissertation at MIT, advised by Jule Charney, on the dynamics of the same monsoon storms that I was now observing from an airplane. To write it, I had relied solely on data collected over land, the only kind available at the time. When I was named the chief scientist of MONEX's Bay of Bengal field experiment, I didn't feel like a kid let loose in a candy store—I felt like a kid who had just been handed the deed to the candy store itself.

But it wasn't just the monsoon research that excited me. I was also hard at work on proving a hypothesis that many in my field dismissed as an impossibility—I believed that, in addition to predicting the weather five and ten days in advance, we could also predict monthly and seasonal averages. Long-term seasonal prediction, as I envisioned it, would not only improve lives around the world but also *save* them. Floods foreseen—famines prevented. No longer would the weather's volatile whims sneak up on society's poorest and most vulnerable, people like the ones I had grown up alongside.

I knew that the reams of data recorded during the Global Weather Experiment would certainly play an important role in this dream.

In July 1979, about one hundred fifty of us—scientists from universities and federal agencies, students, flight crews, project managers, journalists, and archivists—gathered in a Calcutta airport hotel for the month. For two weeks we planned out field experiments and for two more we performed intensive observations, flying our borrowed airplanes every day that an enticing monsoon depression formed. After each day of flights, the scientists gathered in a charmless conference room and studied the map—hand-drawn back then—that the MONEX data had so far created. We'd surmise and theorize about what had happened and why, a warm-up for the long years of analysis that lay ahead. In a little over twelve months, we would gather again to present papers on what MONEX had taught us.

I doubt anyone enjoyed the month more than I did: I was a monsoon-obsessed scientist in his beloved home country, delighted to be reunited with the sights and sounds of India—the familiar rhythms of Hindi songs that filled the city streets, the smell of hot oil and fried bread, the cooling monsoon rains that came down in sheets.

Of course, complex scientific endeavors like the Global Weather Experiment look one way in a planning meeting and quite another on the ground, and there certainly was no shortage of mishaps, tension, and drama during our four weeks in Calcutta.

Most of the time that I wasn't in the airplane, I was plotting out the next day's flight path while watching intently for any pop-up storms.

My fellow scientists and I were eager to be in the sky as much as possible, and I'm afraid that we badgered our pilot more than we should have about getting us up there. Some colleagues had warned me that he could be overly cautious, this former NASA pilot, and within days, his behavior had confirmed that. He told us his flight crew needed rest, the plane needed daily maintenance, conditions were too rough—but these excuses did little to quell our zeal to be airborne. There were only so many weeks that we had the personnel and equipment in place to do our work, and each wasted moment was an underdeveloped forecast, an unwritten dissertation.

There were worse tragedies too. During a lull in our work, some members of our group decided to visit the Calcutta weather department, where recently installed radar systems were gathering all sorts of exciting information. Finding the building among the twisting, sardine-can-narrow streets of the city turned out to be an hours-long challenge, especially as they received wildly inaccurate directions from a number of well-meaning Calcuttans (Indians are much too polite to ever say "I don't know" to a person looking for help). When they finally arrived, my scientist friends were dismayed to find that the radar equipment had been turned off. The Indian meteorologists who greeted them assured them it was for the good of the electronics. "The radar needs to rest for a few hours," they said, confoundingly.

Meanwhile, I was fielding requests from Westerners in our group who wanted to sightsee in a city that wasn't nearly as excited to see them as they were to see it. At the time, West Bengal was under Communist control, and American visitors were immediately flagged as potential spies. The US embassy had counseled traveling scientists to keep a low profile, and I had to constantly warn my colleagues away from late-night excursions and forays into bars.

In fact, on one of the first days of our field experiments, after we flew an airplane into a severe thunderstorm for the good of mankind, the newspapers in Calcutta reported only that we had dropped dropsondes and fuel into the Bay of Bengal. "Americans Pollute Bay," the headlines read.

A lot had changed in the eighteen years between Kennedy's UN speech and the execution of the Global Weather Experiment, but some deficiencies—technological acumen and diplomatic relations among them—had clearly persisted.

And then, of course, there was the matter of the cracked windshield.

After about ten minutes and a couple of nervously eaten mangoes, we finally saw the silhouette of Calcutta appear on the horizon. The fuel streaming from the side of the plane had slowed to a trickle, but my heart was still knocking against my chest. Everyone but Joach remained fearful and silent.

Joach—who had lived for years at the tops of some of the world's tallest peaks studying the behavior of air masses in those terrains, who had flown sailplanes straight into vicious winter storms, who had prepared none other than Alan Shepard for his journey into space—remained exasperatingly serene.

"Have I ever told you about the Gigant?" he called out over the drone of the engine. He had. During World War II, Joach had been flight-testing the aircraft—the world's largest at the time—when it broke apart at twenty-six thousand feet. Now didn't seem like the ideal moment for him to retell the story, but Joach didn't wait for me to say so. "My parachute wouldn't open until I was at maybe two hundred meters. Now, *that* was a close one!"

Out of the corner of my eye, I saw the others exchange nervous glances. I looked at Joach and then nodded quietly toward them, hoping he would get the hint. I knew he was talking about near-misses because he had survived them, and he was telling us that we would survive this one, but the happy ending in Joach's stories was not what stood out to passengers aboard a potentially doomed plane.

"Oh, and in '37, the Rhone Buzzard," he continued, referring to an open-cockpit glider in which he had broken the world altitude record by climbing to twenty-two thousand feet, "My feet went numb, and my fingernails were blue. I was seeing two suns!" Joach let loose a hearty German laugh. "When I woke up, I was drenched in gasoline."

Just then, the pilot (that morning I had thought of him as my grumpy adversary but now I regarded him as my guardian angel) came on the intercom and told us to prepare for landing. The six of us faced forward in our seats, and when the wheels touched the runway, I felt a hundred pounds of worry slide off my shoulders. The voices of several relieved scientists filled the cabin.

A few hours later, we were back in the hotel conference room for the evening discussion. As the weather maps were being drawn up for the evening, Dev Raj slid into the seat beside me to relate a fact that he had learned from a fellow scientist: The windshield of an Electra turboprop plane had seven layers of glass, and while our plane certainly needed maintenance to address the crack, we had likely been no closer to death on that day's flight than we had on any of the others. A few moments later, the pilot himself appeared at the table. To my great bewilderment, he took a newspaper clipping from his pocket and laid it before me. "Shukla's Plea Denied," the headline read. (On further inspection, I saw that the article was about an Indian politician named Shukla who was fighting a court case.) I couldn't help but smile, thinking of how satisfied our pilot must have been to scare the pushy scientists straight. I figured we'd both gotten what we wanted—he'd proved his point, but we had our data.

At the end of our time in Calcutta, there was no grand ceremony or celebration. Now we would spend months in head-down analysis of the data we had gathered. Over the next few years, more than a hundred papers would be written based on the information our instruments had scraped, sponged, and soaked up from the atmosphere, but on my flight home from India, all I knew was that we had done what we set out to do. It was a victory, but like most scientific victories, a quiet and incremental one. Impaired, low-flying airplanes aside, work as a climate scientist could often feel like this, plodding and methodical, a pursuit not necessarily of better answers but of better questions for next time.

OMENS

Since humans began walking upright, we have been striving for control, for a way to transform the chaos of the natural world into some kind of order. Historically, we'd proven to be pretty successful in this endeavor. We had turned heat into fire, wild animals into companions, tangled fields into food. But there was one thing that had eluded us: the weather. Until fairly recently, humans could barely understand what caused the weather, let alone predict it.

We open apps on our phones to see what we should wear and switch to the news to decide if we should make those outdoor plans, so it can be easy to forget that the numbers we receive—the high temperatures, wind speeds, and precipitation chances—are the products of billions of equations solved on supercomputers and millennia of endeavors on the part of obsessive philosophers, adventurers, and scientists.

As early as the seventh century BC, civilization embarked on formalized weather prediction. Some of humanity's earliest recorded weather forecasts were found on ancient Assyrian tablets preserved in a cave in modern-day Iraq. Commissioned by the Assyrian king Ashurbanipal and carved into clay by astrologers and magicians, the predictions were largely based on omens—a halo around the sun, for example, or the shape of the clouds. "If lightning flashes from south to east there will be rain and floods," they warned, and "if the voice of the weather god is heard in the month of Tammuz, the crops will prosper."[2]

For centuries to follow, weather prediction looked a lot like this—hard-won knowledge based on omens, which is an old-fashioned way of saying *empirical observation*. That is, if A often happens before B, then B can be predicted by A. "If the sun is surrounded by a ring there will be rain," and so on. Our ancestors could say *what* happened but not *why* it happened.

By the time Jesus walked the earth, people knew a few of these strategies. "You know how to interpret the appearance of the sky," Jesus says in the Book of Matthew. "When it is evening, you say, 'It will be fair weather; for the sky is red.' And in the morning, 'It will be stormy today, for the sky is red and threatening.'"

Meanwhile, near the end of the Western Han dynasty, the Chinese identified the twenty-four solar terms—particular astronomical or natural events—that made up the year's climate and devised a calendar that provided a tidy schedule for farmers and a slate of weather-associated festivals for everyone else. Instead of a single day in March heralding the coming of spring, its myriad glories were celebrated every couple of weeks: Rain Water, Waking of Insects, Pure Brightness.

Aristotle posited in the fourth century BC that all natural phenomena—everything from earthquakes to rainbows—were the result of cold or hot gases, what he called exhalations. A few hundred years later, Pliny the Elder decided the appearance and disappearance of certain constellations foreshadowed precipitation, and the field of astrometeorology became so widely accepted that in 1492, Christopher Columbus used the positions of heavenly bodies to predict the dates of storms during his voyage to America.

In the seventeenth century, many in England looking for weather predictions turned to the *Shepherd of Banbury's Rules*, a handy guide that contained tips the author had gleaned from years spent afield. If the clouds looked like rocks and towers, the book said, there would be showers. If the clouds were small and round, fair weather was on the way. In 1827, a popular encyclopedia explained that rain was preceded by toothaches, ants clambering over their hills carrying eggs, or candles flaring. (Today we recognize these predictions as crude estimations, but my guess is that many of us have noticed a reclining cow and thought about rain or experienced some level of curiosity regarding what a prognosticating groundhog might have to say about winter's lingering grip.)

In 1858, the Smithsonian Institution in Washington, DC, began displaying a daily nationwide weather map in its lobby, a scientific novelty that quickly became a tourist attraction.[3] It might seem outlandish that something as prosaic as a weather map could become such a spectacle, but then again, is there a topic those of us in the twenty-first century discuss more than weather? I would argue that today, nearly a millennium after humans scrawled weather forecasts on stone tablets, we are *still* fascinated by weather because it mirrors so much of our existence—both awesome and terrifying, repetitive and volatile, inevitable and indecipherable.

It's only in the past hundred years or so that we've begun to figure out the *why* of weather, to move from omens and observations to a more dynamic understanding based on math and physics of what makes the wind blow and the snow fall. By the middle of the twentieth century, new technology such as satellites and supercomputers were revolutionizing the field of meteorology, and the scientists I worked alongside had finally begun to understand what causes weather—and even how to predict it.

Two

In the remote village where I grew up, we knew the weather when it hit us on the head. This was thirty years before and five hundred miles away from where that airplane swung low over the angry Bay of Bengal in July 1979. In Mirdha, a tiny outpost not far from the Nepalese border, there were no radios, no newspapers, no books. Our calendar was the seasons; our clock was the sun.

Our understanding of the weather might have been primitive, but our relationship with it was intimate, deep, and ancient. There was hardly a moment of our lives when the weather was not our most conspicuous companion. Sometimes, it was a joyful host, inviting us to lie in swaying fields under puffy white clouds. Other times, it was a wrathful parent, punishing us with drought or fearsome heat. But, either way the weather was always in charge.

Without sturdy kerosene lamps, we endured stormy evenings in darkness. Depending on the weather's fickle moods, my primary-school education took place beneath the spidery branches of a banyan tree or, when it rained, on the hay-strewn floor of the cowshed. On frigid winter mornings, while the adults did their chores, we children raced outside, placed our backs against the east wall of the house, and lifted our chins to the sky. *Dhoop kha rahe hain*—we were, in the literal translation, eating the sun's rays. During the dry months that preceded the monsoon, we watched for the dust devils—massive walls of dust that blotted out the sun—approaching the village with the roar of a cavalry. Almost everyone in the village rushed inside to escape the storms, which were powerful enough to knock over trees and snatch roofs from houses. But for the young boys, it was a rite of passage to brave the chaos of wind and debris in the relative refuge of the mango grove, where a victory over the *kali andhi*—the black storm—was celebrated with stolen fruit, our theft concealed by thick plumes of dust.

The happiest, most consequential moment of every year was the day the monsoon rains began to fall. In the weeks leading up to the first storm, villagers exchanged predictions like stockbrokers, their guesses based solely on the ephemeral: Evocative breezes. Migrating birds. The smell of something damp. A priest in a nearby village owned a *panchang* (kind of like a Hindu farmers' almanac) that contained its own monsoon forecasts based on the positions of heavenly bodies; sometimes rumors of those predictions made it to Mirdha via a visitor in an oxcart.

The rest of us simply trained our gaze on the sky and waited. Especially the farmers, who stood at the ready; that first drop of rain was like a starting gun, and when it fell, they would race to the fields and sow the rice seeds their livelihoods depended on. Often, there were false alarms, a big storm and then . . . nothing. To fall for this deception was ruinous—no rice for your family, no money to buy more—so many stood with one eye on the clouds, one eye on the most experienced farmer in town.

The wait was torturous. Locusts scavenged what was left of the fields. Flowers slumped over in defeat. The trees rattled their empty branches. The whole world fell silent.

When the skies finally opened, when the first fat kerplunk of rain smacked the chapped red earth, there was only one thing to do. Go outside and get completely, joyfully, gloriously wet! While my friends and I jumped in ever-widening puddles, the world around us changed in an instant. Birdsong blared from every direction and swarms of mosquitoes materialized from thin air. Walking paths and hillsides washed away as mud and muck and mire and swamp seeped across the land. Soon, everything that was brown became green, and seedlings of rice shot up in rows. Our wells once again brimmed with water.

Eventually the novelty of all that rain wore off; endless storms swept through the village until it seemed like our bare feet would never be dry again. And the monsoon we'd waited for, hoped for, prayed for became the guest that overstayed its welcome. Now we waited, hoped, and prayed for it to end. We were eager for the dry season and its holiday festivals.

Worse, though, were the years the monsoon rains did not come no matter how many prayers were offered to the rain gods, no matter how much we watched the sky, willing it to turn black with thunderclouds. *Sukha, sukha* was all anyone said. Dry, so dry. The whole world was thirsty, the ground full of gaping crevasses, desperate for a drink.

Joy and sorrow, pain and comfort—everything depended on the weather, and the weather yielded to no one. It was a god on earth, as powerful as it was inexplicable.

Every year, the monsoon rains—or lack thereof—highlighted the difference between my family and the others in our village. During the active periods of the monsoon, when thunderstorms bellowed across the plains, we huddled together in our home, warm beneath blankets and rugs. However, for the very poor (words that described the majority of villagers), the monsoon rain didn't happen only outside; it percolated through the holes in their thatched roofs, soaking their clothes, the straw they slept on, their children's hair, for months. During the dry years, my family had food reserves and, thanks to my father's job as a schoolteacher, the means to travel to a larger village to buy more. Most villagers couldn't afford to store food for the dry years; they lived hand to mouth. A drought meant skinny neighbors with despairing faces, servants with hunger in their eyes, bony and milk-barren cows.

Our good fortune was mostly the result of luck. We were Brahmins, the highest caste in Indian society. My father and his twin brother were the sixth generation of Shuklas to be born in Mirdha. When they were just eight years old, they were married to two sisters from a Brahmin family in a neighboring village; the story goes that both boys fell asleep during the long religious ceremony. After many years, when his first marriage did not produce a son, my father married a second wife. He was in his mid-thirties, and by then, his hair had turned mostly white, so he dyed it black for the ceremony, reportedly the first person in the village to do such a thing. The elder villagers used to joke about it when I was a child.

We all lived together in a big clay house: my father, his two wives, all my siblings—two sisters, my brother, and me (two other children had died as infants). I was the second-oldest son, born two years after my brother Mahendra. I called my father's first wife "Big Mother" and his second wife simply "Mother." The two women spent most of their time together, gossiping, cleaning up the most recent meal, and preparing the next. Every day we ate chapati or rice and dal; we had okra or eggplant when it was in season or potatoes and onions. As Brahmins, we never ate meat, fish, or eggs. My father's twin brother and his eleven children and many grandchildren lived next door, so our house was always full of family—and the smell of something cooking. While this appears so unfair in hindsight, Mahendra and I, the two sons, ate first and as much as we could; the two daughters ate whatever was left.

As far as I was concerned, my mother wore a halo; it was almost as real to me as the many colored bangles she wore around her wrists. She answered the door to every beggar and readily gave away our food and blankets. (Seeing the joy on my mother's face during her encounters with the beggars instilled in me the idea that helping others was one of the surest paths to happiness.) After each meal, she'd offer the children the skirt of her sari so that we could wipe our faces and hands. The white cotton that fell from her shoulders was always stained orange with turmeric and curry.

My father, by contrast, was quite domineering and strict. He wasn't home often—back then, the men of the family always slept outside (spending too much time with women, even their wives, was seen as unmanly)—but when he was around, my siblings and cousins would scatter like bugs. They feared him.

My father was a schoolteacher in a larger village about three miles away. His monthly salary on top of ten acres of farming land made my family relatively better off than the other villagers, most of whom had no land and slept in huts. When he bought the bicycle that he rode to work every day, it was the first time anyone in Mirdha had seen such a contraption, and a large crowd gathered to admire it. My father was also the only man in the village to own a watch and one of the few who knew how to read. People from miles away brought their letters to our home so that he could read them aloud.

I wanted very much to please my parents; I saved the best mangoes I pilfered from the mango grove for them. While Mahendra was always getting in trouble, I followed the rules, rarely got in fights with my cousins, and never spoke out of turn. When I was eight years old, I volunteered to be the family priest. This position required me to rise early, bathe outside at the well, and light the ghee lamps in the prayer room. Every Brahmin family had a prayer room with shelves filled with figures of deities and portraits of Hindu gods; ours was small and dark, the air thick with the smell of sandalwood.

Every morning, I tiptoed into the prayer room and began to light the tiny ghee lamps before the golden gods perched on the shelf. I always started with Ganesh, who had a man's body and an elephant's head, and then I shuffled down the line of deities, chanting the prayers that came so naturally, it was like I was born knowing them. I moved quickly past Kali, the blue-faced goddess of destruction with her pink tongue outstretched, and lingered longer before Saraswati, the goddess of learning, whose four hands held a rosary, a book, a stringed instrument, and a vessel of water. When all the candles were flickering and the incense was burning, I'd close my eyes and pray as the room around me filled with the strong fragrance. Only when I had finished this ceremony would my mother begin to prepare our breakfast.

My universe was quite happy and very small. I had no concept of how large the world outside the village was, but there were moments when hints of it began to creep into my understanding.

One day when I was young, I was by myself on the dirt path that led out of town. All of a sudden, the ground began to tremble, and a massive gray beast appeared on the horizon. Alone and terrified, I began to sob loudly. From atop his elephant, the handler tried to reassure me. "Just move out of the way," he yelled to where I was crouched in the dirt, but my body seemed to be frozen solid. I had never seen anything so large, so imposing. The elephant came nearer and nearer, its white tusks pointed forward like spears. I cried even harder. At the last minute, the handler was able to steer his animal around me, but it came so close, I could hear the crunching of the rocks beneath its feet. Ever since, elephants have pursued me in my nightmares.

I was introduced to even greater wonders on my first trip to Ballia City, a large town about ten miles south of Mirdha. My family had traveled there in an ox-drawn cart to take part in a ceremony called a *mundana,* a Hindu sacrament in which a young child's hair—thought to represent sins from past lives—is shorn for the first time and offered to the gods. I don't remember whose *mundana* it was, only that many women and children in my family had come to watch the child's wispy black hair be tossed into the Ganges, the holiest river in India. What I *do* remember is the train.

It thundered past us at a railway crossing just outside the city as all the children sat rapt in the back of the cart. The terrible sound it produced, the rattling of every nearby thing—we had witnessed this kind of ferocity only from the sky. When the train's whistle blew, some of the children began to cry. That night, many of us could not sleep as the train's fury echoed in our heads. (Later in the trip, we witnessed an incredible magic trick: someone showed us his new ceiling fan, which he could make move by flipping a switch on the wall!)

I was reminded of the train the day a large truck drove down the dirt path of the village. It was the first time most of the people there had seen an automobile, and many of the children ran after the truck, amazed by the peculiar smell of motor oil and the clouds of dust it left in its wake. It was like the *kali andhi*, except this black storm seemed to come and go at the behest of an ordinary human being.

The man who controlled a mighty elephant; the man who filled a room with wind; the train and truck thundering by with what appeared to be no effort on the part of their drivers—these were my first glimpses of the way that humans could exert power over a natural world that, until then, had seemed untamable.

THE KING OF MONSOONS

In Mirdha, the monsoon wasn't simply one thing. It was *everything*: a governing body, a gift from the gods, a holiday, a common language. Back then, however, we had no idea why it came or why it didn't or that its arrival was heralded by forces many hundreds of miles away from our tiny, dusty village.

Monsoon does not mean "rain." It does not mean "downpour," "deluge," or "torrent." In fact, what makes a monsoon a monsoon has less to do with water and much more to do with wind.

Despite how we use the word in common parlance, the word *monsoon* seems to have originated from the Arabic word *mausam*, which means "season." A monsoon is the *seasonal* reversal in the direction of the prevailing, or strongest, wind. It is this reversal that causes the weather events we typically associate with a monsoon, including months of soaking rain. But there are dry monsoons too.

What on earth could cause the wind to simply turn around? The Earth itself, actually.

The land beneath our houses, office parks, and tennis courts may seem as unchanging as a hunk of plastic, but in reality, it is constantly undergoing its own transformation, heating up or cooling down depending on the season and the tilt of the planet's axis. This change in temperature leads to a change in air pressure, which influences the direction of the wind.

Air is always trying to move from areas of high pressure to areas of low pressure, but its path is never direct, thanks to the rotation of the Earth. It's this movement of air that creates wind, from a lovely sea breeze to the bone-chilling gust of a winter day. Monsoons happen when the land heats up and creates an area of low pressure; this attracts the moisture-laden winds from the ocean, which begin to churn toward that low pressure. As the moist air rises and cools, the water vapor in the air condenses and forms clouds—and the heavy rain we associate with the word *monsoon*.

Three months after the monsoon starts, when the warm season ends, and the land cools faster than the ocean, the winds reverse once again. The winter monsoon takes moisture from the land with it, leaving behind crystalline skies, cold temperatures, and dry weather. A dry monsoon.

Sixty percent of the world's population—billions of people—live in monsoonal regions, areas where drinking water, hydroelectric energy, and agriculture all depend on the arrival and intensity of the annual monsoon. But the most famous monsoon of all, the king of monsoons, is India's summer monsoon.

The intensity of the Indian monsoon is very much the product of unique geography—the large landmass of the Indian subcontinent lying north of the vast Indian Ocean. During the spring months, both the land and the ocean become superheated. As the winds reverse from a southeasterly to southwesterly course, large quantities of water from the sea evaporate and are drawn into the wind, now blowing across India. Eventually, all that moisture-laden air makes its way to the Himalayas, which lie across the northern border of the country like a big brick wall. With nowhere left to go, the air rises and condenses into clouds that produce the months of torrential and relentless rain.

In India, where nearly half the population works in the agricultural sector and the majority of sown lands are irrigated by monsoonal rains,[1] the monsoon has a profound impact on the economic well-being of the state and the individual, so much so that it has been called the country's real finance minister, and during years of monsoon droughts and crop failures, suicides skyrocket.[2]

The Indian monsoon has also soaked into the country's folklore, literature, and art: Paintings that depict anxious people looking skyward. Classical melodies that summon storm clouds. Poems that describe the exquisite agony of being away from a loved one during the joyful monsoon rains. In fact, one of the most common themes of Indian music is the desire to be with one's lover during the rainy season. Many Indian palaces feature a dedicated balcony upon which a nobleman could enjoy both the downpour and his harem of women. Even the Indian notion of beauty—dark hair, eyes bright as lightning—reflects the culture's fondness for stormy weather. In the West, people seek out sunshine and silver linings. In India, there is nothing more hopeful than the merciful shadow of a nimbus cloud blown in by the monsoon wind.

Three

In India—at least, in the India of my childhood—there is a strong superstition that if a traveler sees a cow being milked and glimpses the actual milk itself, his journey will be a safe one and its purpose achieved.

When I was seven years old, my father and I set out for Ballia City, where I would begin the fourth grade at a government school with a very good reputation. During the week, I would be staying at the school's youth hostel, and so, perched on the back of my father's bike, I clutched a small bag with a few of my belongings and a clean pair of kurta, the white cotton pants that look like pajamas that I wore every day.

There was a knot in my stomach. I did not want to go. I wanted to stay in the village with my mother and my many cousins. But I also, perhaps just as badly, wanted to make my father proud. As we bicycled down the dirt path out of the village, I kept my eyes open wide so that the wind would dry my tears.

After only a mile or so, my father very abruptly planted his feet on the dusty ground and stopped the bike. He was quiet. After a few seconds, inexplicably, he turned the bike 180 degrees. Trying to figure out what was blocking our way, I looked down the dirt path. Again, my father spun the bike, and I turned my head, trying to focus even farther down the path in order to identify why we were at an impasse.

I felt the muscles in my father's back tighten. "Why will you not look toward the cow?" he shouted.

Startled, I looked in the direction my father was looking. That's when I saw the lopsided little cowshed that stood by the path. I saw the cow and the farmer milking it. I saw the milk in the bucket.

I was young; I did not know that people believed this sight was a good omen or that, for it to work, one had to behold the scene by accident. Sighing, my father put his feet back on the pedals. We rode the rest of the way to Ballia City in silence.

My father dropped me off at the school, and a week later he came back to the city to check on me. When the supervisor of the hostel—called warden—greeted him at the gate, he had a worried look on his face. He told my father I had been crying every day since he'd left and refusing to eat. He said I might not be ready to live away from home.

My father had very high hopes for me. He wanted me to be successful, and he knew that meant I had to get the best education possible. And *that* meant I needed to leave the village. But, as he told the warden, my mother was just as forlorn as I was. She too had been crying all week and refusing food. Even though it meant I would not go to the top primary school in the region, my father put me on the back of his bicycle and brought me home, where I resumed the normal life of a village child—gathering cow dung in the mornings, playing in the muddy fields during the monsoon, and attending the rudimentary primary school beneath the banyan tree. If I had just seen the lucky milk on my own, perhaps things would have gone differently.

But my father was not about to let a little thing like homesickness stand in the way of my education. He began campaigning for a primary school building in Mirdha. Lessons under the giant old banyan tree were interrupted too often by monsoon rains and wandering cows, he argued. Many of the non-Brahmin families objected to the idea, arguing that education would corrupt their children, who should be out working in the fields. Nevertheless, after about a year of politicking, my father succeeded in getting a small school built just south of the village (its location was chosen to satisfy its opponents, who insisted the village not be downwind of the questionable institution). It had mud walls, two rooms, and a roof that kept them relatively dry in the rain. The last year of my primary school took place in a building. Unfortunately, the building lacked doors, and many mornings we would arrive to find that it had been made a mess of by the jackals, foxes, and nilgai that traipsed through at night.

None of that mattered to me. I loved school. I wanted to be a teacher just like my father. My older brother, Mahendra, had other plans. He wanted to be a wrestler.

In addition to soccer and jousting, wrestling was a major pastime in the village. And while the first two tended to happen haphazardly—players relying on stones for goals and bamboo for lances—wrestling was a more organized affair. Typically, each wrestler represented a village, and their contests would be announced a day or two in advance and spread by word of mouth. These muscular, formidable men would square off in a homemade wrestling ring of tilled soil, wearing nothing but long loincloths, while hundreds of villagers looked on, cheering, jeering, and cringing with each violent maneuver. The wrestlers relied a lot on strength but also on skills, gimmicks, and tricks, not unlike the wrestling matches on TV today.

My brother was strong and fearless and a very good wrestler. Of course my father didn't approve of Mahendra's ambition—or that he ignored his schooling to pursue it—but Mahendra remained determined to be the best wrestler in the region. We were so different, he and I. Whereas Mahendra found my father's expectations constricting, I was happy to adhere to the vision my father had for me. In his tidy designs for my future, I found great comfort. For every season, a plan; for every problem, a solution.

After one year in the new primary school, I began attending the middle school where my father had become the principal, about three miles from the village. I walked to school barefoot (really) and then, three years after that, I went to the nearby high school. My father wanted me to study science so that one day I might be an engineer, but the high school that was within walking distance from the village had courses only in math, Hindi, economics, and Sanskrit, so that was what I took. At the end of tenth grade, when I was twelve years old, it was time for me to take my high-school exam and move on to intermediate college. That was when I learned a very surprising fact about myself: I was not twelve years old at all.

In remote Indian villages like mine, every child was born at home and there was no official method for recording the date of birth. Back then, the first time the birthday of a child had to be documented was the day the child began primary school, when the parent was trusted to provide the correct age. This honor system was quite corrupt, and most parents made their children two or three years younger than they actually were so they appeared advanced for their age. Because no one knew the month or day a child was born, often teachers just wrote down the day the student arrived at primary school as his or her birthday. Almost all the kids I grew up with had "birthdays" in July. My own was July 17, 1946.

In 1958, the Indian government decided to put a stop to this academic fraud and passed a law that said children could not take the tenth-grade high-school exam until they were fourteen. According to my birthday, I would be twelve at the end of tenth grade, and therefore I was too young to take it. Again my father began his politicking.

He went to the board in charge of high-school education and somehow convinced them to change my birth date from July 17, 1946, to July 17, 1944. In the span of a single afternoon, I became two years older.

I passed my high-school exams and my new birthday was stamped onto the diploma, forever sealing my age.

There were still more hurdles to come. My father had not given up his desire that I study science and become an engineer. But because I had never taken a science class in my life, no college would accept me as a science major. No matter—I would simply have to catch up, he decided. He visited the principal (a fellow Brahmin) of a college in Ballia City and they struck a deal: provisional admittance if I passed a comprehensive science test. That day, my father came home with a stack of books, more books than I had ever seen together in my life—the science textbooks for grades six, seven, eight, nine, and ten. He said I would not be wandering around and grazing the cows like I had every other summer but rather studying for my exam, which, somehow, I passed.

Perhaps because my father was so insistent that I study science, the subject came naturally to me. I especially loved physics and delighted in learning about the invisible forces that made things happen, that explained why objects plummeted to earth, why it was more difficult to walk up a hill than down. However, over the next few years of intermediate college in Ballia City, I discovered that chemistry and biology were not for me. The chemistry lab stank like sulfur, and biology students had to dissect live frogs. Thanks to my Brahmin upbringing, our strict vegetarianism, I was clearly not cut out for a career in the life sciences.

At last, it was time for me to leave home. On a steamy July day in 1960, when I was (indeed) sixteen, I stepped off the train in Varanasi, a large city about one hundred miles from Mirdha. I couldn't believe what I saw.

Traffic everywhere. Cars, trucks, rickshaws, cows, all clogging the narrow roads that led through India's oldest, holiest city. In addition to the million or so people who lived there, droves of pilgrims visited Varanasi each day, many of whom had no intention of ever returning home. Hindus believe that dying in Varanasi and being cremated along the River Ganges guarantees a direct flight to heaven, breaking the endless cycle of death and rebirth and delivering the soul at last to salvation. The city is famous for its many thousands of temples as well as its many ghats, steps that lead down to the riverbank, where funeral pyres burn day and night.

I was relieved when the manual rickshaw delivered me to my final destination. Luckily, my university was a few miles away from this madness, from the pervasive smell of exhaust and the persistent screeching of car horns. Banaras Hindu University was founded by Pandit Madan Mohan Malviya, a great scholar, educational reformer, and fellow freedom fighter with Mahatma Gandhi. He was very proud of Hinduism. In spite of repeated requests from Gandhi not to use the word *Hindu* in the name of the university, he kept it. The university's beautiful campus would become my home for four years; it was where I received my bachelor's in physics, math, and geology and my master's in geophysics (my father decided I would not pursue engineering after all when he learned how expensive the national exam was, not to mention how stiff the competition was). Still, I was not immediately happy there. I chose BHU because a relative, A. K. Tiwari, was on the faculty in the math department. A few weeks after I arrived, I visited Dr. Tiwari at his home, a small suitcase in tow. I told him I was going back to Mirdha. Just as it had been when I was a child, the homesickness was overwhelming.

Most of all I missed my mother. I missed her cooking and her affection. I missed my friends in the village and the feeling of comfort I felt there. I was sixteen, and Mirdha felt as indispensable to me as a family member, as central to my identity as my last name.

Dr. Tiwari and his younger brother Brij Kishore Tiwari listened as I described my wish to return home. They said they thought that if I gave it a few more weeks, I would grow to like it in Varanasi. Could I just stick it out for that long and see? they asked. Brij Kishore Tiwari accompanied me back to the hostel, and every few days he stopped by to see if I was okay (or, more likely, to see if I had run away to my village).

It was because of their kindness that I stayed in the university, and BHU started to feel more like home. More than a year later, during one of my infrequent visits to Mirdha, my mother scolded me for being away so long. I hardly even wrote anymore, she said.

It was June in Mirdha, the very beginning of the monsoon season; the air was heavy with humidity. I was home on a break from university, happy to see my family but already missing my life in the city. I had come to like the freedom I had there and to like Varanasi itself, how the streets bloomed with umbrellas during the monsoon rains, the joyful sound of sitars that wafted from the hostel where the music students lived. My homesickness had been assuaged by close friendships I had formed with classmates, other boys from the rural areas. Although we were looked down on by kids from more affluent regions, we were proud of where we had come from. We held illegal poker tournaments late at night, stationing a lookout by the door in case the warden came around.

I also liked when my father came to visit. I loved my father but often felt too nervous to speak to him about anything personal; we didn't have long conversations late into the night in my dorm room or over platters of dal and chapati in a local café. Instead we sat quietly together or else discussed something tedious at length: what I could get him from the canteen; how and where he would like me to make up his bed. During his visits I hoped he felt proud that all the work he had done on my behalf had paid off. I think that he did. In fact, that's why I think he visited in the first place: to see the fruits of his labor.

During that trip home to Mirdha, I slept outside at night with the rest of the men as I had been doing for years. There were dozens of cousins and uncles out there under stars so bright that sometimes it was like sleeping with the light on. My father was not among us; he had long ago built a sleeping porch off the second story of the house. It had a roof to keep the rain at bay and a door on all three sides to let the breeze flow through. Its design was so overtly dangerous that, looking back now, I'm not sure why someone didn't object, but so many things in the village were like that. When you walk barefoot in the dark amid cobras and scorpions, maybe you overlook the smaller, more insidious threats to your safety.

I don't remember who shook me awake the night my father fell; I just remember the frantic words "He fell down" and "He has hurt his back." In truth, my father had broken his spine. He was unconscious when we hoisted him onto a cot for the hour-long journey to Ballia City, the same one he and I had made on his bike all those years before—only now, I was the one lifting his reticent body.

In Ballia City, the doctor made his pronouncement swiftly. We must get my father to a better hospital with more advanced equipment, he said. If we hurried, we could make the morning train to Varanasi. It was decided quickly that I would accompany my father, since I knew my way around Varanasi, while Mahendra stayed behind with my mother. A friend of Mahendra, Sheosagar, volunteered to join me.

Before we got on the train, my mother insisted that we take the jar of ghee—the purified butter that Hindus turn to for luck and healing—she had brought from home. But in the commotion, the jar fell to the ground and shattered. It was a distressing omen; my mother began to sob.

Once we were on the train, as the sky outside turned purple, then orange, then blue, my father briefly surfaced to consciousness a few times, only to groan or shift uncomfortably. Sitting next to the cot we had carried him on, I kept my hand on his arm. I realized I hadn't said a word in hours.

By midmorning, we reached Varanasi. Together, Sheosagar and I lugged my father's cot off the platform and into the busy street, where each of us held a hand aloft to hail a taxi. When at last a dusty yellow car stopped in front of us, we set the cot down, and I placed my hands beneath my father's arms. We were so close to the university hospital, I told myself; just a few more minutes and my father would have the attention of well-qualified doctors. But as I placed his heavy body in the threadbare back seat of the taxi, he shook suddenly. Still in my arms, he took one last sharp breath.

My father died while our taxi idled near the most sacred temple ponds in the holiest city of his faith. Sheosagar and I decided we must take my father's body directly to the bank of the River Ganges for cremation. Within an hour, Dr. A. K. Tiwari and Mr. Brij Kishore Tiwari met us at the riverbank. It was all so impossibly simple; there were shops to buy the wood and priests standing by to offer the last Hindu prayers. As I watched my father's body burn, I wondered for the first time about the nature of life, its random cruelty and senseless tragedies. Just as I had once become two years older in a single afternoon, on the day my father died, I changed very quickly—and for good.

That evening, while we waited for the train back to Ballia City, Sheosagar marveled at my father's luck. Without intending to, he had made it to Varanasi to die. As students and commuters and pilgrims bustled by, my father's soul was en route to heaven—no more rebirths, no more deaths, no more detours.

I did not feel lucky. I felt lost. Without my father's gaze, his guidance, his preternatural ability to locate an answer to every question, I wondered what would become of me. At seventeen, I had never made a single decision on my own.

For thirteen days, we mourned my father in Mirdha with various rites and rituals. In one, my aunts took the glass bangles from my mother's wrist and shattered them. As a widow, she would never again be allowed to wear colorful jewelry or clothes.

When at last I returned to Varanasi, the city that I had once been so enamored of felt ruined. In the shadows of its many temples, I felt the chilling absence of my father. The haze of its endless funeral pyres stung my eyes.

I found solace in my textbooks. In the weeks after my father's death, I began as a master's student in exploration geophysics, a track that we had decided upon after learning I would receive a monthly stipend of two hundred rupees per month *and* be eligible for a job with the Oil and Natural Gas Commission (ONGC) of India on one of their oil rigs. Eventually, I could even be one of their research scientists, a job my father would be especially proud of. He might have been gone, but I could still fulfill the dreams he had had for me. One year into my studies, everything going according to our plan, I sent my application for jobs to the university's employment office.

Around that time, the Indian government began recruiting trainees for the India Meteorological Department. My classmates and I did not pay much attention to the advertisements that ran in the newspapers because our specialization was exploration geophysics; we had taken only one or two classes in meteorology, learning the science behind basic weather phenomena like wind and storms. But unbeknownst to us, the university employment office had forwarded our résumés to the government.

Long before I heard any response to my job applications, a message I was not expecting arrived in the mail. It was a postcard from my mother, auspiciously smudged with turmeric paste. *Your marriage has been fixed,* it said. *Please come home.*

THE REASON FOR THE SEASONS

Until the tumult of my young adulthood—the death of my father, a surprise engagement—very few things ever changed in my life. I imagine the Mirdha I was born into looked very much like the Mirdha my father was born into and his father before that. Many people I knew lived and died in the same square mile of land. Their days followed the same pattern; for men, collect dung for fuel in the morning, graze the cows in the afternoon, gather to gossip while the women baked bread in the evening. The only variations any of us could count on were the seasons. At least in this way, our lives were just like people's lives throughout history and all over the world, even people who had lived long ago and in faraway places.

A contemporary of Bach and Handel, Venetian composer Antonio Vivaldi is considered among the greatest of the Baroque composers. Vivaldi wrote more than five hundred concerti, and the most famous of them all is *Le quattro stagioni—The Four Seasons*.

To our modern ears, these four violin concerti sound as conventional as it gets, but when Vivaldi published them in 1725, they were a revolution. *The Four Seasons* was one of the earliest examples of program music—music designed to render a narrative. Each concerto was named for a season and accompanied by a sonnet, which Vivaldi perhaps wrote himself, that described and reveled in the many wonders of that season.

In the autumn, "cooling breezes fan the pleasant air," and winter "brings its own delights." During the spring, thunderstorms cast "their dark mantle over heaven" while a goatherd and his faithful dog lie sleeping in a "flower-strewn meadow." These narrative elements are reflected in the music; violas mimic barking dogs and the orchestra erupts into a lively thunderstorm.

Vivaldi's *Four Seasons* is one of the most famous pieces of music ever composed. Of course, this is largely thanks to the work's musical innovation and exquisite loveliness. But I would argue these concerti have stood the test of time for another reason: every single person who hears them can relate to their themes. Even the coldest and hottest places on earth have seasons—the snow falls on Death Valley; the Arctic summer sun never sets. It has been suggested that large differences in summer and winter are conducive to creativity; most of the Nobel laureates in the United States received the prize for work they did from the Northeast, where the seasons are pronounced.

The passage of the seasons reminds us that, despite the all-consuming drama of our own existence, time is always moving forward. And when we get very old and look back on our life—on the *seasons* of our lives, as the saying goes—we often find that our most vivid memories are accompanied by a seasonal weather report: the glowing autumn leaves that danced above a college quad, spring rain on a wedding day, snow on the ground when the baby came home from the hospital.

But despite humanity's profound physical and emotional connection with the seasons, a lot of people don't know what causes them. In fact, when students graduating from a very prestigious and expensive East Coast university were asked what caused the seasons, nearly all of them said it was that during the winter, the Earth was farthest from the sun, and during the summer, the Earth was closest. But this is the opposite of what happens! (This answer would have been correct if it was given by Australian students.)

While it is true that Earth's orbit is not a perfect circle, our planet's distance from the sun is not as significant in determining the seasons as the tilt of its axis. In the northern hemisphere, as people shiver through winter, they are actually as close to the sun as they will get all year, but the Earth's axis is tilting away from the sun, meaning that hemisphere receives less of its warming energy. During the northern hemisphere's summer, the north half of the globe is as far from the sun as the Earth's orbit will allow, but the planet's axis is tilted toward the sun. As we already learned, days are warmer than nights because during the day, the energy we receive from the sun is greater that the energy we lose to space, and at night, the opposite is true. Likewise, thanks to the tilt of the planet's axis, in December, January, and February in the northern hemisphere, the energy received from the sun is less than the energy lost to space.

The *astronomical* definition of the seasons is based on the precise times of the equinoxes and solstices, which change each year. However, meteorologists define seasons a little differently than astronomers.

In his book *Notes on the State of Virginia*, Thomas Jefferson provided monthly averages of temperature, rainfall, and wind based on his daily observations in Williamsburg, implicitly describing the summer season as the months of June, July, and August because they were the warmest. Jefferson was perhaps the first modern author to give elaborate reports on the impact of climate on people, animals, and food production. It may have also been the first time someone described *meteorological* seasons.

Just like in Williamsburg, the three-month average of temperature for the whole globe is the highest for June, July, and August, which people in the northern hemisphere, like Jefferson, define as meteorological summer, and the lowest is for December, January, and February, which is that hemisphere's meteorological winter. The remainder of the months form the transitional seasons. (For the southern hemisphere, it is the opposite; winter is June, July, and August, and summer is December, January, and February.)

Meteorological seasons are based on three-month groupings and always begin on the same day (in the northern hemisphere, spring starts on March 1, summer on June 1, fall on September 1, and winter on December 1). They are far more straightforward than their astronomical counterparts (the dates of which vary), making it easier for scientists to calculate and compare the seasonal statistics so useful to agriculture and commerce. The near-perfect regularity of seasons makes them highly predictable (as Shelley wrote, "If Winter comes, can Spring be far behind?") and therefore the challenge for climate science is to predict deviations from the mean seasonal conditions. That is a challenge I took on as one of the missions of my life.

But no matter how you define the seasons, they are inarguably one of the best, most inspiring parts of life on earth. The preeminent Indian poet and philosopher Rabindranath Tagore has composed many songs about seasons, especially about monsoon rains. From the triumphant green buds of spring to the first glassy frost of fall, each season is its own splendid symphony.

Four

I took the postcard from my mother back to my dorm room, sat on my cot, and studied it. She must have dictated it to someone, since she could barely write, but it was definitely her signature in Hindi at the bottom. A deep sense of dread blossomed in my chest. For seventeen years, my life had proceeded as predictably as an assembly line. And now, less than six months after my father had died suddenly, another unwelcome surprise. I did not want to be married; I wanted to be a scientist. I was so wrapped up in my own angst, I didn't even notice that the postcard lacked the most basic detail: Marriage to whom?

I made a quick decision: I would not agree to the wedding. When I went home in a few weeks, I would stand my ground. I even made a contingency plan should my relationship with my family be irreparably harmed by my refusal: I would run away from the village and live with the family of one of my classmates in Varanasi.

As soon as I arrived in the village, my family's campaign began. The bride, my uncles informed me, was a bright student from a well-respected Brahmin family in a neighboring village. According to my uncles, her father held an influential position in the city office. Someone said he had promised to give Mahendra a teaching job in a primary school and fix up my father's old property in Ballia City. On and on my brother and uncles went, listing reasons why I should agree to the marriage. I sat there shaking my head no, planning my new life in Varanasi.

At last, when they saw that nothing else would persuade me, they used their eleventh-hour effort, their nuclear option. My family told me that my father had arranged the marriage before he died, that it had been one of his final wishes. Although it was inconceivable to me that my father would make such a promise, I had no way of knowing if that was true or not. But if there was a chance that this marriage was his plan for me, then I felt compelled to honor it. To say no would be dishonoring all the effort he had put into my life. And so, a month later, I returned to Mirdha for my wedding.

In rural India, a wedding isn't merely a few hours on a Saturday afternoon but multiple ceremonies and rituals that can be spread out over years. The first ceremony that I took part in was called tilak, which is also the name of the mark Hindus wear on their foreheads; in this ritual, the bride's father paints a smudge of sandalwood and turmeric on the groom's forehead. It's also when the bride's family offers a dowry to the groom's family. Today, the tilak ceremony represents the culmination of serious negotiations between two families about the size and nature of the dowry. How much cash in public? How much cash in private? How many appliances and what kinds? Will they be steel or silver? A motorcycle or a car?

It seemed to me that all of Mirdha came to see the dowry my family would receive. I sat on a makeshift stage next to a small sandalwood pyre as the bride's father deposited shiny new rupees into my outstretched palms. Each time he gave me a handful of coins, I dropped them into a copper bucket, making an ostentatious clanging that caused quite a stir among the spectators, many of whom had never seen or heard so much money in their lives. Altogether I received three thousand rupees, about forty dollars in today's money.

It was the largest dowry ever given, as far as anyone in the village could remember, which didn't stop my uncles from grumbling that we should have received more.

The next ceremony was the wedding, which usually takes place on a date chosen by the priests based on the horoscopes of the bride and groom. For this ritual, the groom's family travels to the bride's house accompanied by a band, a dance troupe, a gaggle of guests, and as many elephants, camels, and horses as they can afford. The villagers always talked about the wedding of another Shukla, son of a well-known priest; the groom's father had arranged for more than fifty elephants and fifty-eight horses to accompany his son's wedding party. While my family was well-off by the standards of our tiny village, we were able to afford only a single elephant. The bride's family lined up to welcome our convoy, and they appeared displeased by our underwhelming parade, especially the lone four-legged offering.

While virtually every resident of the two villages (another characteristic of Indian weddings is that *everyone* is invited) enjoyed a rousing party, my bride-to-be and I were summoned to her family's home for the long ceremony, which was performed by two priests beneath freshly cut bamboo trees. For hours they droned on and on in Sanskrit, sometimes prompting me to say a prayer or directing the families to offer more money to the cow-dung deities before us. Occasionally, I heard the revelry outside. Meanwhile, my stomach growled; I felt bored and despondent. Beside me, my bride sat behind a gauzy curtain, her face covered by a sari and a shawl. I wondered what she was thinking, if she was as miserable as I was. I did not see her once during the ceremony. I still did not know her name.

Traditionally, at the very end of the night, the two families meet for dinner and yet another gift exchange, and the groom's family makes one final, friendly demand of the bride's father. This is usually a formality, a chance for everyone to toast the marriage and enjoy the celebration. But Mahendra—technically the patriarch of our family, now that my father was gone—demanded a bicycle, and the bride's family had had enough. They said no. My brother would not back down, and neither would my new father-in-law. Suddenly there were whispers and then a stony silence.

A marriage of a village boy and a village girl was like the marriage of dolls. The bride and bridegroom simply sat there and did what they were told to do. As the bridegroom, I sat there silently and did whatever my family wished me to do. Many of my much older uncles and older cousins were there but they did not intervene. Before I knew it, my family was storming off with me in tow—without dinner or so much as a farewell to anyone. Ironically, while the VIP group returned home without eating, the rest of the wedding party had their dinner outside in the open fields and enjoyed the songs and dances.

Traditionally, I would have gone back to my new wife's house in the morning to finally meet her and her mother. But after the prior night's rift, it was decided that the wedding party would return to Mirdha without delay. Because the bride and I were both so young, the next ritual, called *gawana*, would not take place for five more years; that was when our adult lives together would begin. For now, she would live with her family in Ballia City, and I would return to school.

By the time the letter from the ONGC arrived in Varanasi confirming that I had gotten a job on one of their oil rigs, I was a married man.

Dehradun is a picturesque city nestled in the foothills of the Himalayas. It was one of the British colonialists' favorite hill stations, a high-altitude getaway where they could escape the heat and noise of the crowded cities below. Dehradun is also the headquarters of India's Oil and Natural Gas Corporation (its name changed from commission to corporation), and it was where I went to live in 1965 after I graduated from university.

I was assigned an apartment on a tidy suburban street and a roommate. Kaila was a Punjabi physicist who had also just graduated from his master's program; we were both enrolled in ONGC's oil-prospecting training program. Now, a little more than fifteen years after the country had claimed independence, India needed oil and gas to fuel its industrial development and was building the capacity to do just that. In exchange for our one-year apprenticeship, we had to sign a contract promising we would work for ONGC for at least three years after training was over. If we did not stay, we would be responsible for reimbursing the agency for our training. This was a deal we were happy to make because, while we would be working irregular hours out in the field for many years, eventually we would have the chance to be researchers. Real scientists.

Most of our training took place in the oil fields about an hour outside of Dehradun, where the ground was crumbly and barren. Sometimes the work bus would pick us up in the morning, sometimes just as the sun was setting. Kaila and I would sit in the back seat and split the newspaper between us, trading sections as we finished them. When we got to the worksite, our job was to survey. We drilled holes of varying depths and dropped sensors into them in order to map out the subterranean system and measure the properties of the rocks and water below.

It wasn't the most exciting work but it paid enough for me to send money home to my mother at the end of every month to take care of family expenses, so when a letter arrived from the India Meteorological Department with the date and time of my interview in Delhi, I nearly crumpled it up and put it in the trash. I had never imagined working in the weather department; I had a master's degree in exploration geophysics, not in meteorology. I was twenty, married, and working in the job I had gone to school for. Dehradun had lovely mountain views and mild weather, and in a few years, I supposed, my wife would leave her village to come live there with me. She might even like it. My destiny, such as it was, seemed to be set.

But Kaila, who had also gotten an interview, had an idea. An innocent, impulsive suggestion that, in hindsight, marked one of the most important turning points in my life.

"Come to Delhi anyway," he said.

"With what money?" I protested. I hadn't read the letter all the way through like my roommate had. The interview was at the renowned Union Public Service Commission, where interviews for most important government jobs took place. The commission offered to cover second-class train fare to the interviews in Delhi.

"We'll ride third class and use the leftover money for food," Kaila suggested.

"What about a hotel?"

"We can stay with my parents outside the city. After our interviews, I can show you around. You said you've never been—when else will you get the chance to go for free?"

Kaila was right; I did want to go to Delhi, the cosmopolitan capital of India, to see the iconic sites—the Red Fort, India Gate. I would have been reluctant to travel to such a big city alone, but here was my friend offering to be my host and tour guide. So on a breezy spring day in 1965, Kaila and I took our seats in the third-class train car en route to Delhi.

Unlike all the other candidates pacing the halls of the Union Public Service Commission, I felt altogether relaxed as I waited for my name to be called. My most pressing concern was my reimbursement, which I hoped would be given to me immediately after my interview so that Kaila and I could get directly to sightseeing.

When it was my turn, I walked into a sterile conference room and took my seat in front of the three men who would conduct my interview.

"Hello," I greeted them casually, eager to get the meeting over with.

They each offered me a curt nod and began asking questions. Most of them contained familiar terms, things I had learned in my meteorology course in Varanasi, but some of them were about theories and equations and phenomena I had never heard of.

For the questions I didn't know the answer to, I offered a candid, if not chipper, "I don't know."

The last question they asked was about hail. "If you cut into spherical hail," an older, serious-looking man on the right said, "why does it contain concentric circles?"

"Does it really?" I said, leaning forward. "I have never heard that before!"

Now the three men exchanged curious glances, surprised, no doubt, to hear such a flippant answer from an interviewee.

The hail question intrigued me, and I began to think aloud until I felt like I had stumbled upon the right answer (which was that, as hail forms and goes upward inside a cloud, it gathers layers of ice; when it becomes too heavy for the cloud to hold on to, it falls to earth). When I saw them nodding approvingly, I thanked the men for their time and went directly to pick up my per diem.

Three weeks later, when I was back in Dehradun, a letter arrived from the India Meteorological Department saying I had been selected for a job in the Indian Institute of Tropical Meteorology, an organization I had never heard of, in Poona (now called Pune). IITM was a newly established institute for monsoon research, and I had been offered a job as a junior scientific officer. The letter mentioned that funding from the United Nations Development Programme (on the march after Kennedy's rousing speech) would be sending some of the institute officers abroad for training opportunities. In short, I could be a real scientist, and not years from now, but now.

There was just one problem—the contract I had signed with the Oil and Natural Gas Corporation. I had spent several months in my apprenticeship, and it would cost me a small fortune to back out of it. Still, Kaila, who did not get a job offer from the institute, told me I would be crazy to turn it down, and I knew he was right.

That weekend I took the train back to Mirdha to seek my mother's advice. She gave it to me, and she didn't equivocate. "We did not spend so much on your education so that you could work on an oil rig," she said. "You are too young and too small to be out there at night." She was right that I was young, but that was the least of my concerns.

"The money," I argued.

"I will sell my jewelry. You should be working in the capital. You are not going back to the oil rig."

That night I took a walk around the village, my mind ping-ponging between the options before me. It was early summer, well past time for the monsoon rains to have started, but they had not; the land was still cracked and dry, the fields a dull, dead brown. Earlier, when I greeted our neighbors, I had seen that familiar worried look in their eyes, the threat of a serious drought becoming more real by the day. As we chatted, I could tell their thoughts were elsewhere, working on unsolvable calculations: How would they make money this year to feed their families? How long would the water in the well last? How would they keep the cows alive? I had been away for many years now, an eager participant in India's rapidly modernizing society, but every time I came home, the same crises were plaguing the village. The same crises that had been plaguing the village for centuries, in fact.

In 1877, during the prime of my grandfather's life, more than eight million Indians died in a catastrophic monsoon drought. Maybe parts of his experience—the misery but also the determination—had been written into my DNA. Maybe it explained the voice inside me now, urging me forward.

For a moment, I let myself imagine that I could do work at IITM that would make a difference for people in villages like mine. What if they no longer had to rely on hearsay and omens to plan the most important details of their lives? What if there was some way to figure out when the monsoon would arrive and how long it would last or even just how wet it would be?

That I was in a position to make this choice was entirely due to the tireless efforts of my father, the support of my mother, and the wisdom of the many, many teachers who had influenced my thinking. But the decision about whether or not to take the job was mine alone. It was the first one I ever made for myself.

And so the journey began: a bus from the village to Ballia City, a small train from Ballia City to Varanasi and, a big train from Varanasi to Bombay. In Bombay (now called Mumbai), I waited on the platform for the train to Pune.

⁜

The train from Bombay to Pune was so crowded, I had to push myself and my suitcases through the door as the train was beginning to move. After more than two hours in the sweaty compartment, I pushed my way back out and disembarked at Shivajinagar Station (named for the seventeenth-century warrior Chhatrapati Shivaji).

A five-minute walk from the station—and a five-minute walk from my new job at the Institute of Tropical Meteorology—I found a cheap motel that rented rooms by the month. For the first day or two, the motel was full of transient characters, people who didn't say much when you passed them in the corridor. But after a couple of days, the rooms began filling up with residents like me, young people who had come to Pune for their new jobs. The hotel began to feel almost like a dormitory.

During those first few days in Pune, my feelings swung wildly between excitement and trepidation. Excitement because I would be making five hundred rupees a month, a fortune as far as I was concerned and enough for me to send my mother a sizable portion. Trepidation because I had no idea how or why I had gotten the job or what research I was supposed to do.

Part of me wondered if the interview panel in Delhi had misinterpreted my blasé demeanor for earned confidence. In reality, I felt like I had a meager understanding of meteorology and that my grasp of physics was tenuous at best, as I had not had a science teacher until after I graduated from high school. What had I gotten myself into?

On my first day, my heart faintly pounding, the sky above a cheerful blue, I walked the tree-lined streets of Pune to the institute, located in Ramdurg House, a palatial two-story with large verandas; the building must have once belonged to a very rich person.

As a junior scientist, I had to share my office with another colleague, but the space did come with a few perks. For one, there was a window. For another, there was an assistant. His main duty was to sit on a stool outside the office and wait for instructions from the scientists—to take a file somewhere, for instance, or bring us tea and water. To alert him, we pushed a button on our desks. I am sorry to say that his official government job title was *peon*.

My supervisor was a man named Dr. K. R. Saha, a retired air force captain who liked me immediately. The first day we met, he took one look at my white kurta (in America, this would be like showing up to the office in overalls and a straw hat) and announced that it was "good to have a boy from the village!" My rural roots were also on display when I filled out my employment forms. After checking the box for *married,* I was unable to provide the information for the next line—*name of spouse*. I still did not know the name of my wife. When the office clerk found out about my conundrum, he looked at me with a mixture of disbelief and derision.

When I told Dr. Saha that I had a master's in geophysics, his face lit up. He said my first task would be to read a recent review paper about tropical cyclones written by a Japanese scholar and then give a presentation to my coworkers about it. To get an assignment was a relief (I was way too embarrassed to ask anyone what I should be reading) and also, of course, terrifying.

After work, I went immediately to the library and checked out a glossary of meteorology terms. For the next few days, I spent every waking hour—and many that I should have been sleeping—teaching myself about tropical cyclones, which, I learned, was a generic term that scientists use to describe rotating systems of storms and clouds that originate over tropical water. Hurricanes and typhoons are both tropical cyclones.

When the day came for my presentation, I was anxious but prepared. I stood confidently before my colleagues and described the review paper. I spoke of violent winds around the eye of strong storms, the wall of deep convection around the eye, and the clear sky above the eye. I surprised myself when I could answer all the questions raised by the experienced meteorologists. It was a turning point. The senior scientists who had at first viewed me as a boy from the village began to treat me with respect, and my own confidence blossomed.

Soon I realized that researchers were judged by how many scientific articles they published. Unfortunately, to write a paper, I needed a good scientific question and the technical expertise to perform the necessary calculations. I had neither. So I began to read every paper I could get my hands on and jot down the questions they inspired me to ask. Then I approached some of my senior colleagues to see if they wanted to collaborate with me to answer my questions, hoping that, along the way, I could glean some technical expertise from them.

I was a bit flabbergasted when Dr. Suryanarayana agreed to work with me. Surya, as everyone called him, was the one man at the institute who knew how to use its most powerful computer, the IBM 1620. I suggested that we develop a statistical method to establish the relationship between the current five-day average temperature and pressure over Indian weather stations and the following five-day average temperature and pressure over the same Indian stations. Surya agreed.

This involved analyzing fifty years of data over all available Indian stations, a process meteorologists call "looking forward in a rearview mirror." In many ways, we were following in the footsteps of another monsoon-minded scientist, Gilbert Walker. In 1904, the British government sent the mathematician to India as director general of the India Meteorological Department. His assignment was to figure out what had caused the devastating monsoon droughts of the late nineteenth century, when widespread famine resulted in millions of deaths and made the British worry about the empire's future in the colony. Walker analyzed past data, hoping to find some kind of relationship between one weather variable and the intensity of the monsoon rains. (While Walker was successful in initiating long-range forecasting in India, his forecasts were not very accurate.)

Sixty years later, we were essentially doing the same thing, trying to calculate a linear regression, a way to predict the value of one variable based on the value of another. For instance, if the temperature and pressure in Delhi during a particular week in the past yielded a certain temperature and pressure for the next week, we can predict that a similar temperature and pressure this week will likely produce a similar temperature and pressure next week. The idea was to introduce this new computer-based method to make predictions, and the hope was that if weather forecasters knew the temperature and pressure for the next five days, they could make a better forecast of rainfall.

For many years, this was how scientists predicted the weather—they established a relationship between variables by looking at what had happened in the past. Unfortunately, it's a pretty ineffective method for one simple reason: deviations from seasonal cycles do not repeat themselves.

But at the time, the work Surya and I did together, using massive amounts of weather data and a computer, had never been done in India. Although it was not a new idea, our computer-based method for forecasting was novel in the country. Surya and I published a paper that summarized our results. It was my very first paper, and it got a lot of attention from other scientists at the institute.

To my great surprise, they began to refer to me as "the smart boy from the village."

NUMERICAL WEATHER PREDICTION

The experiment that Surya and I did on the IBM 1620 employed a method called statistical weather prediction, a more sophisticated process than simply identifying associated natural phenomena but not that much different, ultimately. For many years, however, statistical weather prediction was the best that scientists could do. That is, until the introduction of a world-changing technology in the mid-nineteenth century kick-started a slow but steady scientific revolution.

It wasn't any newfangled meteorological device that propelled weather prediction into the modern era but rather the telegraph, invented in 1835. Suddenly, it was possible to collect observations from distant locations in real time, allowing would-be forecasters to discern how weather systems evolved and behaved as they swirled around the globe. The word *forecast* was first used by Vice Admiral Robert FitzRoy, an English naval officer and scientist (not to mention the captain of the HMS *Beagle* during Darwin's legendary voyage), who used the telegraph to advance the study of weather and formalize its prediction as a function of government.

A mariner, FitzRoy was keenly attuned to the role the weather played in military strategy, commercial enterprise, and the safety of his fellow seamen. In 1854, he headed up Great Britain's first meteorological department, the forerunner to today's UK Met Office. In that capacity, he oversaw the creation of a vast observation-collection system, installing barometers at nearly every port in the country, establishing fifteen land-based weather stations, and equipping scores of ships with meteorological instruments. All of the data collected was sent to London via the wondrous new electric telegraph, allowing FitzRoy to create what were at the time the most detailed weather maps in history.

In 1859, after one of the North Atlantic's deadliest shipwrecks—the sinking of the *Royal Charter* off the coast of Wales, which killed four hundred fifty men—FitzRoy was given the authority to use his maps to issue storm warnings. Two years later, he published the world's first "weather forecast" in *The Times* of London, calling for a clear day

Shukla's father, Chandrashekhar Shukla (1897–1962); painting based on a low-quality photo with his fellow teachers in 1960.

Shukla and his mother, Sita Devi Shukla, in their native village in India, circa 1985.

Shukla at his home in the village, preparing to travel with a wedding party by elephant.

Gandhi College, established by Shukla and Anne in Mirdha, Ballia, the village of his birth.

D. R. Sikka, K. Krishnan, J. Shukla, R. Suryanarayana from IITM visiting the National Meteorological Center in Washington, DC, 1967.

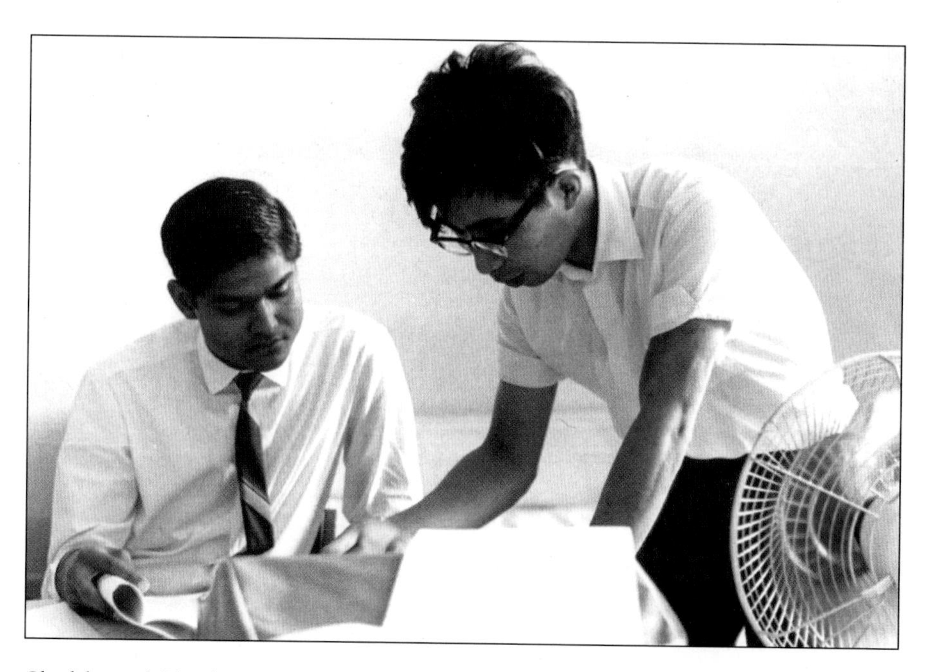

Shukla and Taroh Matsuno at Kyushu University, Kyushu, Japan, 1967.

Shukla and Takashi Nitta at the Japan Meteorological Agency, Tokyo, Japan, 1967.

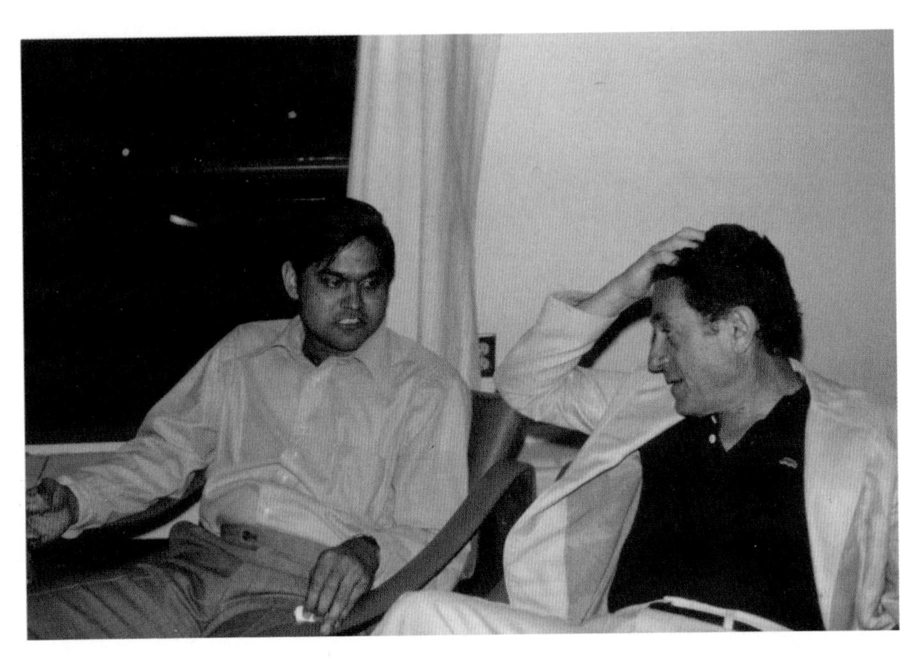

Shukla and Jule Charney in Tang Hall after Shukla's thesis defense at MIT, 1975.

Roger Revelle, D. Lal, Shukla, and other US scientists aboard an Indian Oceanographic Research vessel over the Arabian Sea, circa late 1970s.

Anne and Shukla at GFDL, Princeton, 1979.

Jule Charney visiting Anne and Shukla at their house in Calverton, Maryland, 1978.

Shukla and Abdus Salam, founder, International Center for Theoretical Physics, Trieste, Italy, 1988.

The first US supercomputer (Cray XMP-14) purchased by India for weather forecasting, in the lobby of the National Centre for Medium Range Weather Forecasting (NCMRWF) in New Delhi, February 2024.

Ed Schneider, Shukla, Larry Marx, Dan Paolino at Taj Mahal during a trip to India to implement the first global analysis and forecast system at NCMRWF, 1990.

His Holiness Pope John Paul II (later Saint John Paul) greets the participants of a climate conference at the Pontifical Academy, Vatican, 1986.

Pooja, Chandran, Sonia, Anne, with Shukla, 1992.

in the capital city with highs in the low sixties (a forecast that was pretty spot-on).[1] In short order, this scientific curiosity was syndicated in papers all over England. While they were rarely accurate, FitzRoy's forecasts became wildly popular.

Eventually, however, the novelty wore off and soon people got tired of being caught without an umbrella, literally and figuratively. FitzRoy endured years of scorn and ridicule for his inadequate predictions, eventually dying by suicide in 1865. Today we can look at FitzRoy with much sympathy; the pioneering scientist was doing the best he could with what he had. Weather maps can tell us a lot about what is happening right now, but without knowing and applying the physical laws that govern the atmosphere, you cannot predict the future from those maps.

For the rest of the nineteenth century, forecasters like FitzRoy practiced a version of weather prediction that was a crude amalgamation of data and intuition; they relied on analogs of past weather and instinct when it came to making a guess about what tomorrow might bring.

In 1904, the field once again received a much-needed jolt of innovation. That's when Norwegian physicist Vilhelm Bjerknes published a groundbreaking paper entitled "The Problem of Weather Forecasting Considered from the Viewpoints of Mechanics and Physics." In seven spare pages, Bjerknes proposed the idea of deriving a weather forecast by creating a mathematical model of the atmosphere—in other words, numerical weather prediction, or NWP.

To make an accurate prediction, he argued, scientists required only two things. First, they had to know the initial conditions; in other words, what the weather was like in that very moment. Second, they needed the correct mathematical formulas to describe how those conditions would change over time. Luckily for the article's readers, Bjerknes had a firm scientific basis for those formulas. These primitive equations, as they became known, were based on the laws of motion and thermodynamics, and when applied correctly, they could predict what the weather would do in five minutes, in one day, in one year. The only snag was that in 1904, scientists did not have the observational power to gather an adequate reading of the day's weather, nor was there a man or machine with the capacity to solve the astronomical number of equations necessary to make a useful prediction in time for it to matter.

But that didn't stop scientists from trying.

One of the most valiant efforts came from British mathematician Lewis Fry Richardson, a Quaker who, over the course of many months, managed to produce a retroactive six-hour forecast using the conditions of May 20, 1910, for an area near Munich, Germany. The forecast was inaccurate (it was not his calculations but his chosen equations that were the trouble), but still, it represented a major step forward: the world's first weather forecast derived by direct computation. In 1922, Richardson published *Weather Prediction by Numerical Process;* in it, he described his imagined weather-forecast factory, which was staffed by sixty-four thousand human calculators making and sharing continuous calculations using data supplied by weather balloons spaced every two hundred kilometers around the world. *This* was what it might take to yield an accurate and timely forecast, he wrote.

As it turns out, what the world needed was not tens of thousands of human calculators but one very powerful computer.

The first computer-generated weather forecast was produced in 1950 at the Institute for Advanced Study in Princeton, New Jersey. The effort was led by world-renowned mathematician John von Neumann, who is remembered today as the father of modern computers, and Jule Charney, a name by now you are familiar with. Von Neumann, who thought highly of Richardson's approach, directed the construction of the computer, and Charney provided the meteorological know-how, successfully simplifying Bjerknes's equations to filter out extraneous processes that had plagued Richardson's forecast. It took twenty-four hours to make predictions for twenty-four hours in advance, but the scientific basis for NWP had been firmly established.

This momentous achievement spurred the development of additional weather-prediction models, which were painstakingly created by researchers at scientific institutions around the world—those with access to computers, anyway.

Today, most countries have their own weather models, a badge of prestige but also a matter of national security. All weather-prediction models work in approximately the same way. They divide the atmosphere up into a grid of three-dimensional cells and solve a large number of complex equations in each cell, accounting for variables like temperature, pressure, humidity, wind speed, and a lot more. These equations provide the rate of change of weather in each cell based on the current weather in all the cells. To get the future values of those variables—aka the forecast—the rate of change is multiplied by the length of the forecast and then added to the current value.

Weather forecasts from different models, different countries, and different apps on your phone all vary slightly because there are no universally accepted laws that account for the effect of phenomena like clouds, hail, and squall lines. Each model uses different techniques and spits out different results. You might have realized this while watching coverage of an approaching hurricane. That's when TV meteorologists often compare the European model of the storm's path to the American model.

The European model, developed by scientists from several European nations, is generally considered the most accurate in the world. The American model is no slouch, however, performing eight *quadrillion* calculations per second on one of the fastest supercomputers in the world.

Five

With a little wind in my sails, I started looking around for a compelling research topic. During my first few months at the institute, I heard my colleagues discussing something called NWP, numerical weather prediction. The way they talked about it, I could tell it was the newest and most exciting thing happening in the field. But I had no idea what NWP was, and I was too scared to ask anyone. While I was gaining some respect at the institute, I still felt insecure about accidentally revealing how much I didn't know. There were only two senior scientists in the country actively involved in NWP research, one in Pune and one in Delhi, and their reputations bordered on godlike. There was no way I was going to get access to superstars like that.

So once again, I returned to the library and started reading everything I could get my hands on, this time about NWP. And I learned that it involved taking a simple mathematical model of the atmosphere and solving an immense number of equations to figure out what the weather patterns would do next. I had no idea how those models had come to be or which scientists had discovered the right equations or even how to solve those equations, but I figured if I found step-by-step instructions for one of the simplest equations, I could slowly teach myself how NWP was done.

As it turned out, a Japanese researcher had recently published a paper giving a recipe, essentially, for how to make a short-term prediction of a large-scale weather pattern using what meteorologists refer to as a barotropic model—a model in which the atmosphere is represented by only one layer (by comparison, today's models can divide the atmosphere into hundreds of layers). For weeks, I made it my singular mission to understand this paper, and I followed his process to a tee. I even did the computer programming myself.

The computer that Surya and I had done our experiments on was made up of two processors, one that looked like a blocky gray desk with rows of lights and buttons and the other the size and shape of a vending machine. At the time, the IBM 1620 was among the most powerful computers at the institute and yet it processed data millions of times slower than even our cell phones today. Like all computers back then, it lacked a monitor and keyboard; it was coded by punch card, a stiff piece of paper full of seemingly random rectangular holes, and printed its results on long reams of perforated paper.

To write a program, a programmer would first spell out the code on forms called coding sheets. He would then convert these sheets into punch cards using a keypunch machine, which looked a lot like a typewriter. Often, a second person would punch the very same cards and compare the first set to the second to ensure there were no errors. Last, the cards would be inserted into a card-reading machine that would translate all those holes into electronic signals. Entering data for these programs happened the same way.

Writing code, punching thousands of cards, feeding the cards into the machine—it was time-consuming, tedious work. But the result allowed a scientist like me to make calculations much, much faster than humanly possible. It also meant I got to spend my days in the only air-conditioned room in the institute (the cold air kept the computer from overheating).

In following the Japanese researcher's recipe, I taught myself how to solve a mathematical equation that gave the rate of change of weather from one place to the other. First, I punched in the actual numerical values of weather (temperature or wind velocity) at different geographical points. Then the computer calculated the difference between the two values and made predictions based on the rate of change, first for five minutes, then for ten, until it had made a twenty-four-hour forecast. In truth, the calculations were very simple, equations I could have solved by hand. But the computer could do it much faster, millions of times faster.

Once I had mastered that recipe, I decided to see if I could write my own. I set my mind to predicting the wind patterns over India in a twenty-four-hour period using the computer skills I had learned, the barotropic model, and the weather observations of the day—that is, the initial conditions. I got to work on the long process of punching cards and feeding them into the computer.

Within a year of becoming a junior scientific officer, I had done it. I had succeeded in writing a computer program on the 1620 that could solve the equations necessary to make a twenty-four-hour forecast of wind patterns over India. No one cared how accurate the forecast actually was so long as it resembled a realistic atmospheric flow pattern. Suddenly I began receiving lots of attention from my colleagues, who stopped by my office to see the boy from the village who had created such a sensation. I was even invited to a conference to present my results.

I began working on more programs, this time on an even more powerful computer located about three hours away in Bombay at the Tata Institute of Fundamental Research, an organization that carried out research in various scientific fields. For this endeavor, I devised a "remote computing system": Our assistant took the train to Bombay each day, boxes of punch cards in hand. In Bombay, he submitted those cards, picked up the outputs from the day before, and returned to Pune on the evening train.

As I took on more complex projects, I also began the work I had come to Pune to do—analyzing past observations of monsoon rainfall and wind patterns to study the onset and intensity of the summer monsoon. But no patterns emerged in the data, and I remained puzzled as to why monsoon rainfall was so different from one year to the next, especially because the sun, the land, the oceans, the hills and valleys, and all major astronomical factors remained identical. Although I was publishing papers and speaking at conferences, I couldn't shake the feeling that I didn't really know what I was doing—that I didn't fully understand the physics or the dynamics of the atmosphere. I hoped my confidence would grow with my knowledge, but progress on both fronts felt like slow going.

In 1966, about a year and a half after I arrived in Pune, I began itching to leave. I had come to the institute feeling inspired to help people like the ones in my village, but so far all I had done, it seemed, was punch thousands upon thousands of punch cards. I wanted to improve lives and wasn't convinced that a daylong forecast that took months to produce was going to do anyone much good. To make matters worse, my mother and my older brother, while proud I had secured a government job, didn't seem all that impressed with my work on weather prediction.

That's when a colleague who lived a floor above me at the hotel let it slip he was preparing to compete for a position in the Indian Administrative Service, the most respected and feared branch of the Indian government. It recruited civil servants for government positions—such as judges, district magistrates, and the heads of agencies—around the country. IAS officers had prestige and they had power, the kind that would make my family proud. But as I listened to my friend talk about his dream of becoming an IAS officer, I was more struck by the opportunity he would have to make real and profound change. I decided I would compete too.

I was twenty-two years old, and my ambition was matched only by my naïveté. After I applied for IAS but well before I heard back, I told one of my bosses—Dr. Pisharoty, the director of IITM and one of the most respected meteorologists in the country—that I had applied for another job and would likely leave if I got selected. It was late one evening at the office. The sky outside was a deep purple. I'll never forget the look of scorn on his face or the silence he let settle between us before turning and walking away

A month later I was informed that I was the recipient of a United Nations Development Programme fellowship for the institute and that I would be spending the next eight months in the United States and Japan. Dr. Pisharoty had nominated me; perhaps he had done it to entice me to stay at the institute, or perhaps he had nominated me before I told him I might leave and now he deeply regretted it. Either way, he did not speak to me again until I accepted the fellowship.

There were four fellows traveling in January 1967 from India; our first stop was the National Weather Service in Washington, DC, where we were a bit gobsmacked by the technology on display. We were exposed to the most up-to-date scientific and technical advances in meteorology but learned quickly there would be no formal training. I learned a lot just by regularly speaking with a visitor from Japan, Takashi Nitta, who became my lifelong friend.

We rented an unfurnished apartment to share for the four months we were in DC. We slept on the floor and made improvised meals on an ancient gas stove. When we weren't working, we went sightseeing. A visit to the White House—virtually wide open to the public back then—was a chance for me to don my very first suit and tie. On other days, we were tourists in the mall, wandering around department stores that seemed to sell everything we could think of; we went to grocery stores filled with an unbelievable number of choices. Even the roads were a marvel—such clean and wide lanes full of shiny "imported cars," as we called them in India.

When our four months were up, two of us, Dev Raj Sikka (who would one day fly in MONEX planes with me) and I, were sent to the recently formed National Hurricane Center in Miami, where we stayed in a dorm on the University of Miami campus. In 1967, after a decade of carnage, the United States was working hard to develop a hurricane warning network. In 1954 alone, an astonishing sixteen tropical storms blew across the Atlantic, seven of them full-on hurricanes. Three major hurricanes—Alice, Barbara, and Hazel—were responsible for some one thousand deaths, including that of legendary forecaster Grady Norton, who died from a stroke after tracking Hazel's path for more than twelve hours despite warnings from his doctor about his dangerously high blood pressure.

I was excited to visit the National Hurricane Center, as tropical cyclones had become something of an obsession of mine since my first lecture at IITM, but it turned out that there wasn't much for us to do there either. Our time in Miami ended up being less science experiment and more social experiment.

My first few months in America were surprising, to say the least. I found myself very fascinated by the work culture in the United States, how employees all arrived at the office on time (in India, time is more of a relative concept) and spoke endlessly about the weekend. "What did you do this weekend?" they would ask me. "Ready for the weekend?" They always seemed underwhelmed at my responses. I didn't yet understand that the weekend was for special things—a day trip or purposeful rest. To me, it was just another day of the week.

Meanwhile, life on a college campus provided me with a crash course on what it was like to grow up in this country. I could hardly believe it when I saw young men and women kissing in the halls of the dormitory or lying on blankets in the school's grassy quad, their legs intertwined like pretzels. I thought that it was outright immoral for young students to kiss each other and act as if they were married. When I developed some ailments, especially a persistent pain in my stomach, the manager of the training program in Washington authorized me to see a doctor. After a thorough medical examination showed nothing physically wrong with me, the doctor concluded I was suffering from anxiety. Culture shock, he said. The answer to my anxiety was simply patience; in time, I became used to the different cultural standards and practices of the United States.

After we spent six months in America, it was time to leave Miami for Japan. America had been a bit of a bust, but Japan promised high excitement. I was going to work at the Japan Meteorological Agency under Dr. Kanzaburo Gambo, a respected scientist who was attempting to incorporate computers and numerical weather prediction into the country's meteorological efforts.

Dr. Gambo had spent time at the Institute for Advanced Study at Princeton and met a meteorologist named Jule Charney. He showed me some work this Charney had done on vertical coupling—a fancy way to say interaction—between the upper and lower layers of the tropical atmosphere. According to Charney, there was little interaction between those layers, which placed serious limitations on weather prediction in the tropics, since weather happens only when air moves up or down (think of warm, moist air that rises to make rain).

Gambo encouraged me to work on Charney's idea to see if I could prove or disprove his claims and gave me access to a simple two-layer model of the atmosphere that he had created on the agency's IBM 704. He directed me to run a couple of very basic simulations on the computer using his model: one experiment as a control, and the other where I changed the vertical structure of the temperature and moisture in the atmosphere, which meteorologists refer to as "vertical stability."

The model runs were the quickest part of the process and were bookended by long hours of studying—first, Gambo's research on his model, then the reams of paper the computer spit out. In the late 1970s, running a weather-prediction model produced maps populated by rows of different numbers and letters that the researcher had to carefully analyze to understand what changes the experiment had produced in the computer's version of the atmosphere.

For two months in Tokyo I worked on this problem, and I

continued to plug away at it when I returned to Pune. Finally, I had my results. My experiments had proven the opposite of Charney's conclusions and shown that when the vertical structure of temperature and moisture in the atmosphere resembled their values during India's summer monsoon, there was a significant level of vertical coupling—allowing lower-level moisture to rise, condense, and make rain—which should enable meteorologists to make more accurate predictions.

Gambo encouraged me to submit my findings to the International Symposium on Numerical Weather Prediction, the biggest, most prestigious conference in the field. It was a long shot that my paper would get accepted or, if it was, that it would be chosen for an oral presentation, and a longer shot still that IITM would fund my travel to the event, which happened to be taking place in Tokyo the next year, 1968. When I found out that the two senior scientists, famous in the country for their expertise in NWP, were vying to go to the conference, I thought for sure my chances were slim.

I got very lucky. Not only was my paper selected for oral presentation, but the director general of the India Meteorological Department, who, rumor had it, was not a fan of either senior scientist, decided that I should receive the funding to go. Perhaps he felt like he had no choice, since neither of the other scientists had even submitted a paper; maybe he was impressed that such a junior scientist had the gumption to submit a paper to such a major conference. Indeed, when I found out I would be returning to Tokyo and presenting at the conference, I was filled with two emotions: disbelief and dread.

It seemed everyone in the world was at the 1968 International Symposium on Numerical Weather Prediction. Hundreds of scholars milled around the ornate ballrooms of the hotel, crystal chandeliers hanging over their heads. Unlike the smaller conferences I had attended in India, this one was full of journalists gripping steno pads and waiters carrying trays of hors d'oeuvres. Men in suits smoked cigarettes while loudly debating the merits of recent journal articles. I was certainly a long way from the dusty cow paths of Mirdha.

The scene had me feeling quite overwhelmed, so I tried to stay focused on what I had come to do: present my paper on vertical coupling and raise objections to the paper by Jule Charney that Dr. Gambo had shown me. Because I was so new to this field, I did not know that Charney was the most famous meteorologist in the world, that among scientists who studied weather, he was a veritable celebrity. I certainly didn't know that he was a pioneer of numerical weather prediction who had collaborated closely with the renowned mathematician John von Neumann to produce the world's first computer-derived weather forecast or that they'd worked on it down the hall from Albert Einstein's office at the Institute for Advanced Study when Robert Oppenheimer was its director. But a few hours before my presentation, it dawned on me that I should get a glimpse of the person whose work I was there to criticize, so I asked one of the organizers to point out this Jule Charney to me.

My stomach sank when I saw him, a distinguished-looking man in his fifties with a mop of thick hair, deep-set eyes, and the easy smile of a movie star. That day, I noticed that every time Charney stood to speak, the room fell to an eager hush, silent save for a battery of cameras snapping away. At the coffee break before my talk, he was surrounded by a throng of admirers. Meanwhile, I was approached by journalists who wanted to interview me about how it felt to be the youngest person at this symposium. By the time I was called to the podium, my heart was in my shoes.

I barreled through my presentation so quickly, the chairman twice had to interrupt and tell me to slow down. Only when I finished the talk and waited for questions did I finally look up. I was relieved to see there were no raised hands. But then one hand did emerge from the sea of spectators before me. Charney's.

"I have four questions," he said, a friendly smile on his face.

For a few moments, Charney and I went back and forth about his work and mine. In hindsight, I can't imagine how I even had the courage to speak. I did not win the debate, but I managed, I think, to present a cogent argument. Of course Charney had the last word. "Your paper proves my hypothesis," he said matter-of-factly.

The moment that forever changed my life happened at the end of the session. As most attendees filed out of the conference room, Charney walked up to my table. And while several people lined up to speak with the famous meteorologist, he spent the next several minutes with me, explaining some of the new work he had done on tropical atmospheres. I was so surprised, I could hardly concentrate on his words. All I could think was that the work I had done at the IITM—where my colleagues were already predicting I would take over as director general one day—paled in comparison to the brilliant science happening in other parts of the world. It wasn't just Charney; it was every talk I had heard that day. Suddenly, the limits of my own trajectory in India seemed perfectly visible. Charney asked me to follow him to his hotel room so that he could give me a preprint of his latest paper. The people waiting to speak to him followed us too, and when we got to Charney's suite, we found it covered in papers and books. Later I would learn that Charney was in the midst of planning the Global Weather Experiment; that's why he was buried in paperwork and in especially high demand among his fellow scientists that day. All those people trailing us down the corridors of the hotel were the luminaries of the weather world.

By the time we said goodbye, I felt something new taking root inside me—that confidence, that certainty I had been lacking. I had stumbled into the conference by accident, but I left with a new sense of purpose.

When I returned home from Tokyo, I immediately wrote to Charney to thank him for spending his coveted time with a junior scientific officer from India. I also said that I would very much like to come study with him at MIT.

Charney didn't reply, but a few weeks later, when I was in Mirdha visiting my family, the postman parked his bicycle in front of our home and loudly announced that he was delivering a letter from America, the first one he had ever held. Soon a crowd gathered to see the envelope, the dignified red and blue stripes lining the edges.

It was from the chairman of the department of meteorology at MIT, a Dr. Norman Phillips. Inside was an application to graduate school.

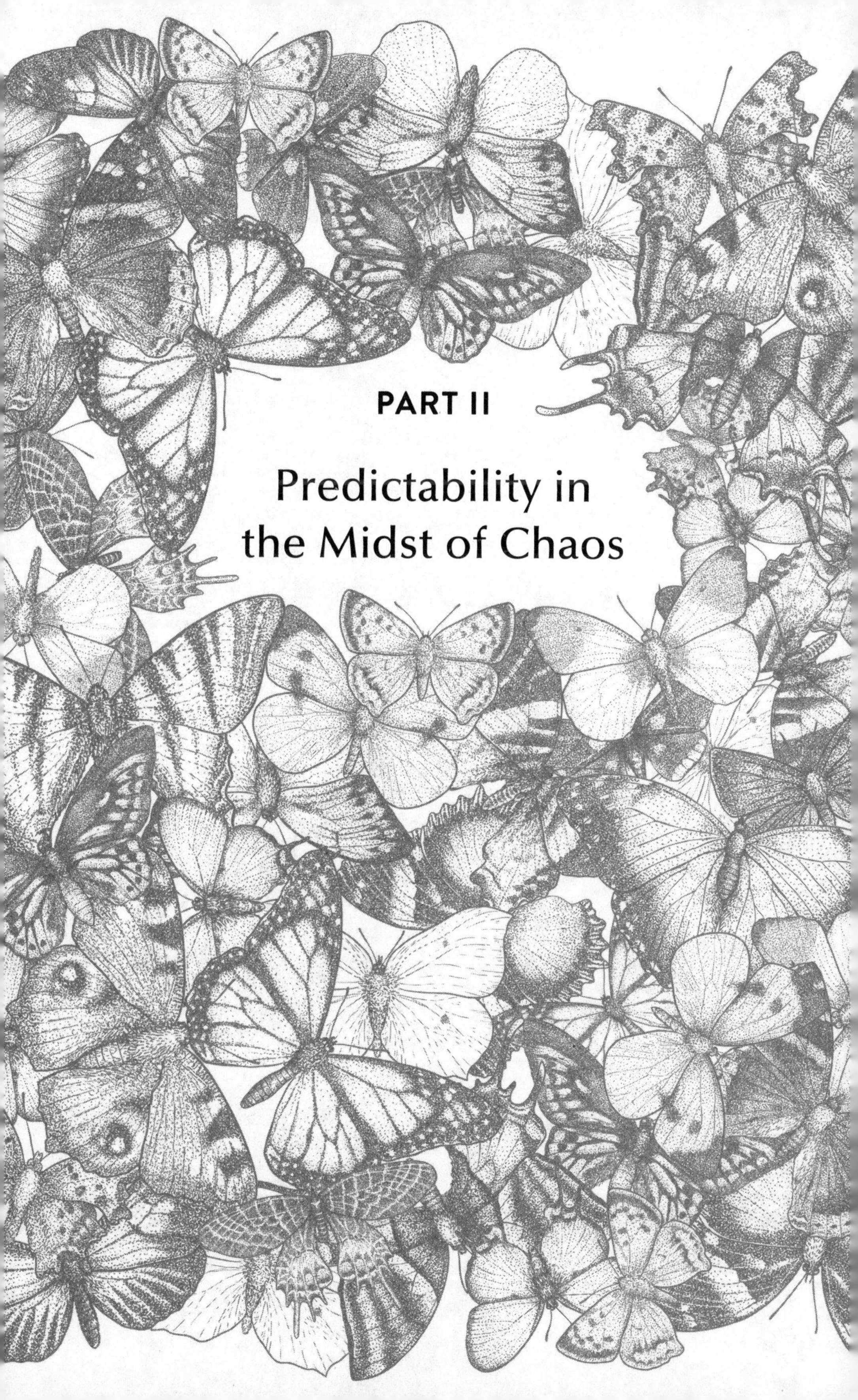

PART II

Predictability in the Midst of Chaos

Six

I returned to America (on a Fulbright travel grant, thankfully, which eased the financial burden), which triggered another case of culture shock, but by far the most difficult transition was my tumble down the ladder from a respected scientific officer with his own peon to a lowly graduate student who had to cook his own dinner. While I was working in Pune, I submitted some papers I'd written to Banaras Hindu University and was awarded a Ph.D. I thought I had finished my schooling, but now I was back on the other side of the classroom. At MIT I was paid a small stipend, but it was nothing compared to what I had earned working for the institute. At twenty-six, I felt like I was starting all over again.

Of course I was grateful, and a little stunned by my own luck. Somehow, I had landed in a hotbed of meteorology, the place where the discipline was hitting its peak—in terms of both progress and tension. The scientists I met during my first weeks in Cambridge, Massachusetts, were pushing the boundaries of weather prediction while simultaneously slamming up against its barriers. On the one hand, satellites and supercomputers were revolutionizing the field, providing scientists with more data and more complex models than they had ever thought feasible. Suddenly, for example, it was possible to ascertain the air temperature at virtually any point in the atmosphere and then plug that data into a computer to see how it might affect tomorrow's precipitation, whereas before satellite technology, scientists were able to collect data only from locations where they could place physical instrumentation. On the other hand, new theories were emerging about the inherent constraints of that knowledge, theories that seemed to snuff out any hope that we humans would ever be able to peer far into the future.

Nowhere was this push and pull more pronounced than between my two advisers, Jule Charney and Edward Lorenz. Giants in the field, these men would change my thinking—and my life—forever.

Jule Charney was born in San Francisco on New Year's Day, 1917. His parents were Russian immigrants and avowed socialists who exposed their son to books, music, and frank political discussions at the dinner table. When Charney was fourteen years old, he came across a calculus textbook on a relative's bookshelf. As he flipped through its pages, the teenager realized he could already solve some of the equations. A lifelong interest in math and science was born.

Charney studied physics and math at UCLA, and after earning his B.A., he stayed to get a graduate degree in physics. In 1940 he met Jorgen Holmboe, an acolyte of Vilhelm Bjerknes, the scientist who famously proposed that weather could be predicted using only initial conditions and a set of mathematical equations. Holmboe invited Charney to join UCLA's new meteorology department. At the time, very few schools offered such a program, although interest in meteorology was booming thanks to the expansion of the military in the run-up to World War II, when an accurate weather forecast was just as useful to the war effort as a tank or a gun.

At the time, the director of UCLA's weather program was

Bjerknes's son Jacob Bjerknes, who had made a name for himself as a support meteorologist aboard the first airship to reach the North Pole (riding alongside legendary Arctic explorer Roald Amundsen). In 1919, the younger Bjerknes had introduced the idea that air masses and the tumultuous boundaries between them—which he named fronts because they reminded him of World War I battlefields—were significant drivers of weather disturbances.

Among these brilliant visionaries, Charney thrived. In 1945, he completed his doctoral dissertation on the instability of airflows in the middle latitudes of the atmosphere, a project that required a staggering amount of hand calculations, employing a level of mathematics theretofore unseen in meteorology. In 1946, on his way to Oslo to do a postdoc, Charney stopped off at the University of Chicago to meet the world's foremost meteorologist, Carl-Gustaf Rossby. Rossby was the discoverer of what are now called Rossby waves—planetary waves in the atmosphere and oceans created by the rotation of the Earth—and the only meteorologist ever to grace the cover of *Time* magazine (still!). Rossby convinced Charney to put off his fellowship for a year and study with him in Chicago. During that time, he arranged for Charney to attend a meeting at Princeton's Institute for Advanced Study, where he met John von Neumann, who was working to build a supercomputer. Von Neumann believed the computer would play an important role in weather prediction and weather control.

In 1947, Charney began his fellowship in Oslo, and there, inspired by the paradigm-altering technology on the horizon, he got to work refining the equations that Vilhelm Bjerknes had suggested back in 1904 and those that Lewis Fry Richardson had used to make his retroactive six-hour forecast in 1922. In short order, Charney ascertained that by getting rid of some irrelevant noise—sound and gravity waves—a more accurate forecast could be produced. The set of equations that Charney formulated are today called the quasi-geostrophic model and are considered one of the most important achievements in modern meteorology.

On any planet, wind, rotation, and the pressure gradient (the difference in atmospheric pressure from one location to another) are strongly related to one another. A horizontal wind that perfectly balances the pressure-gradient force and the force of rotation is called a geostrophic wind. A horizontal wind that *nearly* balances but doesn't quite is called a quasi-geostrophic wind. That little imbalance is a critical component of storms and other weather disturbances.

In his 1948 paper on this work, Charney wrote: "The motion of large-scale atmospheric disturbances is governed by the laws of conservation of potential temperature and absolute potential vorticity, and by the conditions that the horizontal velocity be quasi-geostrophic and the pressure quasi-hydrostatic." Fifty years later, the famous American meteorologist Norman Phillips called those words "the most effective meteorological statement of this century."[1] Effective enough, in any case, for von Neumann to invite Charney to head the meteorology group at the institute in Princeton.

Charney and his wife took up residence in institute housing, where they became close friends of the Oppenheimers, so intimately intertwined were war, weather, and technology back then. By the spring of 1950, his model was ready for a test drive. Because von Neumann's computer at the institute wasn't quite finished, Charney and his team traveled to Aberdeen Proving Ground in Maryland to attempt a twenty-four-hour forecast on the US Army's ENIAC computer. It took one hundred thousand punch cards, one million calculations, and a full twenty-four hours of processing for the computer to produce the one-day forecast, but the forecast was sound. (Charney sent a copy to Richardson in England, although the lifelong pacifist had abandoned numerical weather prediction after his work was used to support wartime efforts. "An enormous scientific advance," the retired meteorologist wrote back.)

In 1952, von Neumann's computer was finally complete, giving Charney access to a more complex model (it divided the atmosphere into three layers instead of just one) capable of producing longer and more accurate forecasts. When Charney ran a forecast that successfully predicted the Thanksgiving cyclone of 1950 (an enormous blizzard, also called "the Great Appalachian Storm," that unloaded up to sixty inches of snow in parts of West Virginia), the institute—a place where scientific breakthroughs were quite common with scholars like Einstein in residence—hosted a large celebration to honor the momentous achievement of a computer's short-range forecast of a storm that had occurred years ago. (It was reported that Oppenheimer attended the celebration but didn't appear overly enthusiastic about it.)

Now von Neumann and Charney turned their attention to promoting these new methods. They helped establish the Joint Numerical Weather Prediction Unit, a collaboration between the US Weather Bureau, the air force, the navy, and NWP groups around the world. They also encouraged the establishment of a special unit in the weather bureau that eventually became the Geophysical Fluid Dynamics Laboratory of the National Oceanic and Atmospheric Administration in Princeton.

At thirty-eight, Charney had completely changed the course of modern meteorology, and he began looking for a university appointment. Administrators at the Massachusetts Institute of Technology expressed interest in appointing him to their esteemed meteorology department and asked what they might do to sweeten the deal. Charney replied that one thing that would lure him to Boston was the promotion of Edward Lorenz, a research scientist at MIT whose work Charney followed closely and respected mightily.

Edward Lorenz was born just a few months after Jule Charney but on the other side of the country, in West Hartford, Connecticut. While arriving in Cambridge, Massachusetts, was a long and winding road for Charney, for Lorenz it was almost a family inheritance; his maternal grandfather had started the chemical engineering program at MIT, and his father had also attended the university, setting a track-and-field record for the two-mile race when he was there.

As a child, Lorenz loved games—especially chess—maps, music, and astronomy. He was also enamored of numbers from a very young age; he recited the address of every house he and his mother passed when she pulled him around the neighborhood in a wagon. After he learned his times tables, the budding scientist took a keen interest in square and cube roots and could soon list every perfect square between 1 and 10,000.

Lorenz earned degrees in mathematics from Dartmouth and Harvard, where one of his advisers was George David Birkhoff, the preeminent American mathematician and a world-renowned expert in dynamical systems. Before Lorenz could finish his doctorate, however, World War II broke out and Lorenz enrolled in the US Army Air Corps to train as a weather forecaster. The special master's program he attended was held at MIT and in it, Lorenz quickly noticed the wide chasm that existed between meteorology and forecasting. The former concerned a scientific understanding of the atmosphere; the latter involved studying past sequences of weather maps and using statistical relationships and intuition to posit the future. Theory and practice were still two very different things.

After training, Lorenz was sent to Hawaii and then Saipan to help set up a weather-forecasting operation in support of Allied bombing raids against Japan. But Lorenz and his colleagues' efforts were hamstrung by the military's lack of upper-air weather observations; most were taken by flight crews themselves. These shortcomings were on full display in 1944 when Admiral William Halsey sailed his fleet directly into a category 2 hurricane churning in the Philippine Sea. Three ships and 790 lives were lost in the preventable tragedy.

When the war ended, Lorenz decided not to finish his Ph.D. in math at Harvard but rather to pursue meteorology at MIT, and in 1948 he wrote a doctoral dissertation that applied fluid dynamical equations to understanding atmospheric motions. Doctorate under his belt, Lorenz signed on to be a research scientist on a project to study the general circulation of the atmosphere and statistical forecasting. (This work would culminate in a landmark monograph, "The Nature and Theory of the General Circulation of the Atmosphere," which remains one of the definitive texts on the subject.) During this time, he visited other meteorology labs, including the one at the Institute for Advanced Study, where he met Jule Charney and John von Neumann.

While working on the statistical-forecasting project, Lorenz, a lifelong mathematician, wanted to test the validity of its basic premise: that it was possible to predict future conditions of the atmosphere by using past data. He created his own model using a Royal McBee Corporation's LGP-30 (an early-model desktop computer than ran just a bit faster than a handheld calculator) and isolated twelve equations to simulate the motion of the atmosphere. Ultimately, he demonstrated that it was *not* possible to make accurate predictions of the future no matter how much information one had about the past. As he would say in a 1960 lecture in Tokyo, the calculations showed signs of "sensitive dependence on initial conditions"—words that foreshadowed his most famous contribution to science.

One day in 1961, Lorenz was working on this model when he decided he would like to see a certain solution in greater detail. Instead of starting the long program over again, he chose to begin in the middle by typing in the numbers the computer had generated about halfway through its last run. Then he stepped out to get a cup of coffee. When he returned about an hour later, he was surprised to see the computer had generated a vastly different solution than it had originally. At first, he thought he was having software problems, since the computer should have come up with the same answer when given the same data. But then he noticed that the numbers for the new solution were identical to those of the old solution for the first few steps of the program; after that, the data gradually diverged until the two solutions were so different, they might as well have been randomly chosen.

Quickly, Lorenz identified the problem: The number he had typed in from the printout went to three decimal places, but the number stored in the computer was accurate to six decimal places. Even though the difference in the data was less than one one-thousandth of a point, this tiny error produced significantly divergent results. Lorenz realized that if the atmosphere behaved the same way, then long-term weather prediction was an impossibility, given that scientists could never fully capture completely accurate initial conditions. The tiny errors that were inadvertently fed into the models would grow exponentially, distorting all the other figures in the model, rendering the forecast useless after just a week or so. Meteorologists might be able to measure the humidity or pressure to the fifth decimal place, but without the sixth, the prediction that computers generated would eventually spit out the wrong calculations. Edward Lorenz had discovered chaos theory, otherwise known as the butterfly effect.

My two advisers could not have been more different. Jule Charney was the star of every party; Edward Lorenz hovered near the perimeter of any social function, if he attended at all. Lorenz seemed to say only what was exactly necessary, while Charney loved to talk, loved to tell his graduate students stories from his days at the Institute for Advanced Study in Princeton.

One evening during my first year at MIT, I went to a small wine-and-cheese party hosted by the department. As usual, the grad students, myself included, formed a circle around Charney and hung on his every word.

"I better stop with the wine or I might start telling the truth," he joked, his thick eyebrows arched in mischief. Of course he didn't stop. Instead, he recounted the day that John von Neumann suggested Charney walk down the hall and introduce himself to Albert Einstein. Dr. Einstein would be interested in the work being done on weather prediction using the laws of physics, von Neumann insisted. Charney admitted that the idea of meeting Einstein made him very nervous. (As Charney said this, we all glanced at one another; we felt the same way about Charney as Charney had felt about Einstein!) When my adviser finally worked up the nerve to go meet the great scientist, he was quite deflated to learn Einstein had no idea who he was—he even confused him with a different scientist.

Another time, Charney told us a story starring von Neumann himself. According to Charney, von Neumann had seriously considered proposing that the US military detonate an atom bomb in the eye of a hurricane on the theory that the energy of the bomb might decrease the intensity of the storm and reduce its destructive power; it would be a test of von Neumann's ideas of weather control. Charney and other weather experts were skeptical about the validity of the hypothesis; they were also deeply concerned that, given the very high-level connections von Neumann had in the defense industry, he might actually succeed in persuading the military to do it. Eventually, they were able to convince von Neumann the idea was a bad one.

To us, Charney represented all the hope, excitement, and even occasional glamour of our burgeoning field. Lorenz was the perfect foil for Charney, a stoic genius who much preferred the lab to a cocktail party. Even their theories were like opposite sides of a coin. Charney's work seemed to be saying, *Anything is possible!* while Lorenz's work replied, *Not so fast!*

After devoting the previous four years in India to monsoon research, I found Lorenz's declaration that long-term weather prediction was impossible more than depressing. After all, I had accepted my first job in meteorology and had come to MIT with the intention of helping people. And helping people meant predicting the weather, especially the monsoon, as far out as possible so that governments and citizens would be able to make well-informed decisions about their economic and physical well-being. If meteorologists had already reached the limit of our forecasting potential, what was the point in carrying on? I wondered.

However, by this point in my life, I was beginning to feel a sense of self-confidence. I couldn't quite explain it, but if a village boy like me could make it to the halls of MIT, perhaps other unlikely things could happen. I wasn't delusional enough to think that I would have some kind of epiphany, some brilliant flash of inspiration; I had heard about the eureka moments of exceptional scientists like Newton, Ramanujan, and Einstein himself, but I wasn't a great believer in so-called genius. And I was exceedingly aware of my mathematical and technical deficiencies. However, I had inherited a bit of my father's unflappable optimism, his belief that every obstacle, whether geographic, bureaucratic, or architectural (the lack of an entire school building), could be overcome, and that, hard work, and a healthy intuition had taken me farther than probably even he had thought possible. I wasn't going to stop now.

Another reason for my newfound confidence was Lorenz himself. I hadn't heard the name Edward Lorenz before I came to MIT, and the first time I saw him, I was taking my seat in his first-semester course on dynamic meteorology. By the end of that class, I knew this man was undoubtedly the greatest teacher I had ever had. I wasn't the only one who thought so; Lorenz won the teaching award so many years in row that the meteorology department decided to simply stop awarding it.

Every class was a revelation. Despite Lorenz's monotone delivery, every sentence was elegant and so brilliant, it was like he was reading from a well-honed, well-rehearsed script. (In fact, we all took such obsessive comprehensive notes that Antonio Moura considered turning his into a book on dynamic meteorology.) Each day, I left the classroom in a daze, dumbfounded by all I had learned and all that I had not known. Finally, I was getting the education I had lacked.

So when Charney found out I had a doctorate and offered me a chance to be his postdoc—a prestigious job that came with a massive pay raise, not to mention a boost back up the ladder—I was torn because it meant giving up my classes. On the one hand, the chance to work under the meteorologist who had essentially transformed numerical weather prediction from a futuristic pipe dream to a practical application was something most grad students would kill for. On the other, I would lose an opportunity to rectify my spotty understanding of the atmosphere and the laws of physics that governed it.

Just three years after my chance meeting with the most famous meteorologist in the world, I turned down the opportunity to be his postdoc and the money that would come with it. I wanted to take my time, to grow as a scientist. I wanted to sit among these brilliant students and scholars and absorb all the knowledge I could. If my plan was to figure out how to forward Charney's work in prediction despite Lorenz's grim prognosis, I was going to need it. I told Charney, I would like to remain a graduate student.

With some trepidation, I told my two mentors that my goal was to demonstrate the predictability of monthly and seasonal averages. I was delighted when both of them, without reservation, encouraged me to proceed.

BUTTERFLIES AND SEAGULLS (CHAOS BY ANY OTHER NAME)

In May of 1996, Edward Lorenz stopped by my home in Maryland for a visit. He had just received an honorary degree from Penn State, and President Clinton had been the commencement speaker. Sitting at my dining-room table, his hands around a glass of water, Lorenz told me that when he was introduced to Clinton, the former president had remarked, "There is lots of chaos in my life."

You don't have to be very science-minded to notice the terms *butterfly effect* and *chaos theory* are bandied about in conversation, mentioned on television shows, and employed by advertisements more than six decades after Lorenz first coined them. Like President Clinton, we all have our own meaning of these terms based on our lived experiences.

You also know that by missing a bus, you could miss a flight, miss an interview, miss the career of your dreams. This helps illustrate another term typically used in a discussion of chaos: the *butterfly effect*. People's lives, just like the weather, exhibit a "sensitive dependence on initial conditions."

But where did that phrase come from?

It is generally believed that the term originated in a lecture that Lorenz gave in 1972 in Washington, DC, entitled "Predictability: Does the Flap of a Butterfly's Wings in Brazil Set Off a Tornado in Texas?" He delivered the lecture long after the science behind it had been published (in his 1963 papers "Deterministic Nonperiodic Flow" and "Predictability of Hydrodynamic Flow").

But the original title of Lorenz's lecture did not include the word *butterfly*. In fact, Lorenz's original title used a completely different animal: "One Flap of a Seagull's Wings Would Be Enough to Alter the Course of the Weather Forever." That day in my dining room, my thirteen-year-old daughter, Sonia, asked Lorenz about chaos theory, and he again used the example of a seagull's wings to explain it to her.

So when did that seagull morph into a butterfly? At that Washington, DC, meeting, Lorenz projected a figure in which he had plotted the values of two of the variables for a fixed value of the third variable in his simple, three-variable nonlinear model to mimic the properties of the atmosphere. The trajectories went around and around without ever intersecting. This property—of solutions that neither repeated themselves nor converged to a steady state—was unprecedented and previously unknown to mathematicians.

When Philip Merilees, a fellow scientist and the person who organized the DC meeting, looked at the results projected on the screen, he saw the shape of a butterfly. Without consulting Lorenz (because he could not reach him), Merilees changed the animal in the title from a seagull to a butterfly. The rest is history.

Lorenz's discovery of chaos theory and the butterfly effect had serious implications for the long-range forecasting of day-to-day weather. One person who understood the profound impact of Lorenz's work was Jule Charney, who, in the years that followed, launched a global study (in preparation for the Global Weather Experiment) to estimate the limit of weather prediction using the most sophisticated and complex global models of the atmosphere available. Charney wanted to make sure that his proposal to improve weather forecasting had a sound scientific basis. To be safe, he titled the resulting report "The Feasibility of a Global Observation and Analysis Experiment."

In 1981, Lorenz visited the European Centre for Medium-Range Weather Forecasts, the undisputed leader in numerical weather prediction. There, Lorenz made new estimates of the rate at which errors would grow using the actual real-time forecasts of the ECMWF model. In a 1982 paper, Lorenz concluded that the reason five- and ten-day forecasts had significantly improved during the previous decades was that one-day forecasts had significantly improved. That is, scientists' ability to assimilate current observations to describe initial conditions had gotten much better.

On April 17, 2004, more than twenty years after Dr. Lorenz's visit to the mecca of weather forecasting and forty years after his seminal papers on chaos and predictability, I had a telephone conversation with my former adviser that he agreed to let me record. I asked him—perhaps not as bluntly—if his work on chaos in the 1960s had been *too* pessimistic, his prognosis for weather forecasting perhaps *too* grim.

Lorenz was silent for a few moments before he spoke. "There was a long period when the satellite data was not helping much, and now it really is. I was surprised when this first happened, and I don't know how long it can keep up," he admitted. "I actually listen to local five-day forecasts and put some stock into them; I never would have done that twenty years ago."

Now my former mentor was ready to update his prescription for better prediction, even in the face of chaos. "The only way you can make [the] five-day forecast better is to make the two-day forecast better."

Later in that phone conversation I asked Lorenz if he would be willing to write a letter to the president urging sustained long-term investment in observation, modeling, and research. To my surprise, he said yes, especially if I could get other scientists to sign it. Ultimately, like so many ideas born in rap sessions between friends, I was not able to get the project off the ground.

Regardless, thanks to Lorenz's original contributions, we now know that some days our forecasts will be much better than other days, but we cannot know which ones in advance. That's why the current procedure, called ensemble forecasting pioneered by my good friend Tim Palmer at ECMWF, is to make twenty to fifty forecasts from the same day's initial conditions by disturbing the initial conditions just a bit—like shaking up a jar of butterflies. If the resulting forecasts vary greatly, the meteorologist knows there's still a lot of uncertainty about what will happen. If many of the forecasts look similar, he or she can feel more confident to make predictions. This is why, when you check the weather forecast for your town, you see a prediction—and then the *probability* that the prediction will come true.

Our quest to manage chaos continues.

Seven

I was happy at MIT, but India was never far from my mind. The deeper I got into my studies, the more determined I became to return home and, somehow, make a difference there. I had begun to read books on social and economic issues in India and I had frequent discussions about Indian rural society with my classmates. I was also taking courses at MIT on political revolutions and, later, on population dynamics; I was even studying the different congressional districts near my village to figure out where I would run for parliament when I got back. Not long after I arrived in Cambridge, my friends came up with a nickname for me: the Senator.

I had been in the United States for more than a year, and I hadn't been able to visit India once. Getting there was such a long and expensive journey, requiring the kind of time and money a graduate student doesn't have. But I daydreamed about India often, about my mother's cooking, the lovely blur of green rice paddies I saw on the train to Ballia City.

Finally, in the fall of 1972, more than a year and a half after I had left, I had the chance to go home, albeit for a sad reason. An Indian physics student at MIT had died after inadvertently touching a 2,000-volt laser, and I was asked to escort his body back to his family in eastern Uttar Pradesh. With my airfare paid for and my time away excused, I would be able to visit Mirdha, although just for one day. Despite my bleak task, I was excited to get back.

But as the plane descended over the agricultural fields outside of the capital city, my excitement evaporated. It was early autumn, and everything should have been verdant and bright, every river and lake filled to the brim. Instead, dry creek beds snaked across the desolate landscape. The trees were barren, the air hazy with dust. I could tell immediately that my homeland was once again suffering from severe drought.

Things were worse in Mirdha, where even my family was short of food, something that had rarely happened during my childhood. As I looked around our struggling village, my thoughts were far away, in the MIT classrooms I had left behind. Surely a phenomenon this extreme and widespread had been heralded somehow. Nature could not be so cruel, I thought, as not to offer us a way to anticipate its life-sustaining variations.

With only hours to visit with my family, I didn't have much time to dwell on the atmospheric physics and dynamics ricocheting around my head. But when my train pulled away in the early morning light the next day, I felt a renewed resolve. I would finish my Ph.D. and waste no time in returning to India.

I had been at MIT for only four semesters when Charney decided to take a yearlong sabbatical in the United Kingdom and Israel. To make sure my academic development wasn't waylaid by his absence, the chair of the department, Norm Phillips, began to look for another institution where I might study while Charney was gone. Phillips called on a good friend in Princeton, Joseph Smagorinsky, who was then the director of the Geophysical Fluid Dynamics Laboratory of NOAA.

One day, Phillips rushed into my office, visibly excited, and told me that Smagorinsky remembered me! Apparently when I was working in Pune, I had attended one of Smagorinsky's lectures and approached him afterward to ask questions. That was enough to get me a one-year appointment as a visiting student at the lab. I would be working under Syukuro Manabe, or Suki, as I came to call him. Months earlier, I had met Suki in the baggage claim office of a Russian airport; both of us were on our way to a monsoon conference in Yerevan, and we had arrived in the Russian airport only to find that our suitcases had not. This was where my future mentor imparted the first of many lessons he would teach me: Always pack the presentation slides in your carry-on.

When I learned that I would spend the year studying under Suki, I was thrilled. Everyone in the field knew that the Japanese scientist had developed the most high-fidelity climate model in the world at Princeton, and now I would have the chance to work on it. I eagerly reported to the Geophysical Fluid Dynamics Laboratory in the summer of 1973.

Suki and his wife, Noko, were immediate friends. A short, slight man, Suki had friendly eyes, a giant gap-toothed smile, and a hearty laugh completely disproportionate to his size. In the lecture hall, he moved with the finesse and excitement of a stage actor; the students loved him. But despite his playful demeanor, he was a formidable genius, and Suki had earned the respect of his fellow meteorologists by creating the very first three-dimensional model of the atmosphere.[1]

This model was developed on the IBM 7030, an astonishingly powerful (for the time) computer that was nicknamed Stretch. Designed at the behest of the Pentagon to simulate the effects of a hydrogen bomb, Stretch weighed thirty-five tons and occupied a space larger than a single-family home. Altogether, nine of the machines had been built; one was installed in the National Security Agency, one at Los Alamos National Lab, and one at the US Weather Bureau, where Suki had begun work in 1958.

In 1966, Suki used a model to run a deceptively simple calculation designed to investigate the relationship between greenhouse gases and the temperature of the Earth's atmosphere. He found that in his model, increasing the concentration of carbon dioxide in the atmosphere from three parts per ten thousand to six parts per ten thousand increased the surface temperature of the Earth by a staggering 4 degrees Fahrenheit (Suki was the first researcher to model anthropogenic global warming, and it earned the Japanese scientist a Nobel Prize in Physics many decades later).

Given the power of his model and my new mentor's reputation, when Suki asked me what I wanted to research during my time in Princeton, I knew my answer had better be a good one. I told him that I wasn't interested in investigating this newfangled global warming—I wanted to predict Indian monsoon rainfall. At this, Suki lit up. He told me about some calculations he had done with his model to study what would happen to Earth's climate if all the mountains in the world were flattened. When he flattened the Himalayas, he said, the monsoon had pretty much disappeared. *This* was the kind of problem I wanted to work on.

In fact, I had come to Princeton with the seed of a similar idea, one that had been planted by Henry Stommel, a legendary MIT oceanographer who studied the Atlantic Gulf Stream. One day, knowing that I had a special interest in monsoon dynamics, Stommel had stopped by my office to show me some new data he thought might be useful to my research. On the Somali coast, he pointed out, the surface temperature of the Arabian Sea could plummet to a surprising 16 degrees Celsius. Perhaps, he suggested, this had some bearing on the Somali jet, the southwesterly wind that blows across the Arabian Sea and brings with it the onset of monsoon rains over India.

Immediately intrigued by this idea, I began analyzing the available data to see if there was a correlation between the surface temperature of the Arabian Sea and the rainfall over India in the same year. Eventually, a pattern did seem to emerge. When the sea was colder, India was dry. When the sea was warmer, India was wet and green. The explanation was simple: Warmer seas meant more water evaporating into the air, which in turn meant more rain. But no one had ever demonstrated this trend using a numerical model of the global atmosphere.

I told Dr. Manabe I wanted to run a series of experiments with his model in which I drastically altered the temperature of the Arabian Sea to see what impact it would have on monsoon rainfall. Since I wasn't well-versed on Suki's supercomputer, one of his researchers helped me run the experiments so I could concentrate on analyzing the results. Over the next few weeks, I created a hypothetical pattern of sea-surface temperature over the Arabian sea in which the ocean was very cold near the African coast, as Stommel had shown me, and then the cold water spread all the way up to India. I purposely used large numbers—2 and 3 degrees Celsius colder near the coast and gradually less cold farther out to sea—to drown out the noise of the model, to make certain the sea surface's temperature was the variable dictating the outcome. But when I think of it now, I realize I was spitballing. No one I knew had ever done a numerical experiment using a global model by changing ocean temperature over a limited area, especially over the Arabian Sea.

In fact, many of my colleagues in meteorology weren't convinced that the effects of modest changes in ocean temperatures could be simulated by a global model, and the early models reflected this belief. Perhaps this was a bit of wishful thinking, since our grasp of ocean dynamics lagged many years behind our understanding of the atmosphere. Of course this stood to reason, as the planet's massive oceans presented an equally massive challenge: they possessed so many boundaries, with so much happening under the surface (literally and metaphorically). The data was nearly impossible to get. That's why the models I worked on at the beginning of my career generally had a fixed sea-surface temperature. And because large bodies of water heat up and cool down quite slowly compared to the atmosphere, meteorologists' short-term predictions didn't suffer too much.

But as my results began to come in at Princeton, I could see right away that the two were in fact intimately connected. For instance, when we entered colder ocean-surface temperature values over the Arabian Sea, the India in our model was plunged into a devastating drought. While these numbers were clearly exaggerations that had never been observed, they demonstrated that my burgeoning hypothesis had legs.

Perhaps there *were* large-scale oceanic and atmospheric conditions that would help us make long-range predictions. I wasn't entirely sure what it all meant, but it was a very promising start.

Even Jule Charney was impressed. He began to tell many of his colleagues about my work, and soon I was receiving correspondence from and meeting scientists I greatly admired, including Francis Bretherton, a British mathematician, and Jerome Namias, one of the leading researchers in long-range forecasting. Namias even asked if I would conduct a similar experiment to see if the Pacific Ocean temperature affected the climate over North America. Unfortunately, my time on Suki's powerful supercomputer was up.

My results at Princeton were a big deal because it was the first time anyone had used a global atmospheric model to demonstrate the relationship between sea-surface temperature changes and monsoon rainfall. But to me, they were an even bigger deal because of the bearing they might have on the people in my village.

At MIT, I took courses on dynamic meteorology, learning about the laws of physics that govern the atmosphere; on physical meteorology, learning about the physical processes of the atmosphere such as radiation and convection; and on synoptic meteorology, learning about the weather patterns we see on weather maps. *These* were the building blocks of knowledge that would have given confidence to any scientist—they certainly did for me. These courses and the brilliant instructors teaching them allowed me to finally understand why weather happens as well as how to predict it.

However, they did not give any clues about why monsoon droughts persist for months and seasons. Still, if we could understand and predict the weather, I began to believe, surely we could do the same with seasonal climate; we just hadn't figured out how yet. By the end of the coursework at MIT, I was nurturing the nerve to think that *I* could be the one to figure it out.

But first, like every graduate student before me, I had to write a thesis. Although I was eager to clear this last hurdle and return home, I was stymied by what topic to choose. By now, it was clear to me that the limitations of weather prediction, the butterfly effect, and chaos theory were the entrenched paradigms in our field, and it would be difficult to propose my audacious ambition—seasonal prediction. I knew I didn't have the mathematical or technical abilities to pull this off just yet (if pulling it off was even possible). Although Lorenz had never said so, his powerful discoveries had led the research community to conclude that seasonal prediction was all but hopeless and that an extension of numerical weather prediction to a seasonal timescale just wasn't in the cards. Now was not the moment to challenge the paradigm.

What I really wanted was for my adviser to simply assign me a topic for my thesis research, the way my father had guided the first sixteen years of my education. I thought that since Charney was funding my education at MIT, he would tell me what to do. In the weeks leading up to my qualifying examinations, I kept badgering Charney, asking him what I should write my thesis about. But every time I did, he would say something vague or cryptic like "It is difficult to come up with a good science question, isn't it?" At first, I wondered if Charney had run out of new ideas. If Jule Charney couldn't define a good science question, how would I? But soon I realized Charney was pushing me to think independently, to find my own thesis topic and pursue the research that interested me. It was time to stand on my own.

Seasonal prediction was off the table for now, but I still wanted to study monsoon rainfall forecasting. In a fit of desperation, I turned once again to my mentor—this time to his old scholarship. After I reread Charney's papers, it occurred to me that just as his groundbreaking thesis research provided a credible theory for why there are winter storms in the extratropical atmosphere, maybe I could propose a theory for the origins of monsoon storms—what scientists call monsoon depressions—during the summer monsoon season. These depressions typically form over the Bay of Bengal and then churn northwest and produce copious rainfall over India. While Charney considered only the vertical shear of the winter westerlies in the atmosphere, I could extend it further by considering both the vertical *and* horizontal shears of the summer monsoon winds.

Vertical shear is the difference between wind velocities at two separate levels in the atmosphere. The upper levels have a fast-blowing jet stream—winds from west to east, called westerlies—while near the surface of the planet, slower winds blow. It is this strong vertical wind shear that is responsible for storms over North America. It occurred to me that I could consider vertical shear for the monsoon region and go one step further by including the horizontal wind shear—the difference in wind speed at two latitudes at the same vertical level—as well.

Now that I had a legitimate topic, Charney was willing to speak with me about my thesis research. He liked the idea but thought that I should do more. He suggested that in addition to vertical and horizontal shear, I should include the latent heat of condensation (the amount of heat released when water vapor condenses into liquid droplets) in my model, and not just any formulation of the latent heat of condensation but the most complex formulation, recently proposed by Japanese climate scientist Akio Arakawa.

For the kind of research Charney was suggesting, I needed to take a nonlinear equation—one that couldn't be solved using simple algebra—and make it linear; that is, solvable.

Nonlinear relationships are ones in which the variables don't demonstrate a constant or predictable dependence on each other. Take, for instance, the relationship between the time we spend at work and our overall happiness. Research has shown that the more people work, the more they feel they have a purpose, and the happier they are—up to a point. After a certain threshold, happiness plummets. Working eighty hours a week doesn't make people feel extra-purposeful, it makes them extra-stressed and tired, thanks to other variables (for example, the need to sleep). Relationships between meteorological variables are like that too—they change depending on circumstances, meaning other variables at play in the atmosphere.

In order for scientists to linearize an equation, they have to do some difficult math equations to make the variables match up in a predictable way. (Nonlinear equations can be "solved" only by the brute force of computers, and even then, only within limits. Think of a weather model—the equations it's programmed with to predict future weather never get totally resolved; the scientist simply runs out of time or the computer runs out of capacity.)

Incorporating Arakawa's formulation into my model was a great idea, but it required me to do this very complicated math by hand in an attempt to linearize a spectrum of clouds of different heights and widths to determine how much rain and heat they produce. If that sounds difficult, it's because it is. My immediate reaction was that Charney's suggestion was impossible. I decided to try anyway.

Luckily, Charney's office and my office faced each other, so if he was in his office (which wasn't too often), I could simply lean into the doorway and ask a question. One day, after struggling for hours with an equation, I looked over and, exasperated, said, "Professor Charney, Arakawa's clouds cannot be linearized."

My adviser glared at me for a second, a look that I interpreted as disappointment. "Everything can be linearized," he said, and he got up and walked away. This was apparently all the advice he was going to give me on the topic. I kept at it and soon realized that I just needed to write two more equations about the way the diameter and the depth of the cloud changed with time. Finally, I was ready to solve these equations using a very fast computer available at the NASA institute in New York that had a remote connection with MIT. Incredibly, the solutions the computer produced were reasonably consistent with the scale and propagation of observed monsoon depressions over India. I could now write my thesis and justifiably claim that I had a theory of monsoon depressions.

But there was one more obstacle I needed to overcome. A few days after I gave the draft of my thesis to Charney, he called me into his office. I found him sitting in there with a strange, irritated look on his face. My thesis was on his desk. "Do you speak German?" he asked.

I was confused. "No, Dr. Charney," I said. "Why?"

"Because your grammar and syntax reads like German," he said, tapping his finger on my thesis.

"Well, I don't have time to learn English now," I retorted. I turned on my heel and headed for my own office. Once there, I slumped down at my desk and closed my eyes. It is a testament to the stress I must have been feeling that I remember my angst that day. Charney's criticism made me feel hopeless, like all my hard work was for naught. I would go back to India without a Ph.D., an embarrassment and a failure. Drowning in my own sorrow, I hardly noticed Charney come in.

"You know," he said, "there are some really distinguished scientists in our field who couldn't write English at the beginning of their careers." It was, for Charney, as good as an apology, an assurance that I would graduate after all. I did require the generous help of my friend and former roommate Venkataramanaiah Krishnamurthy, who typed the thesis and corrected my English.

It didn't take long for my thesis, especially on the heels of my paper on the influence of the Arabian Sea on monsoon rainfall, to make me a minor celebrity—a monsoon celebrity—in India. It was the first time anyone had developed a theoretical explanation for monsoon storms, and more scientists began expressing hope that predicting monsoon droughts could be on the horizon.

Soon after I published my results, several research groups in India launched projects on monsoon depressions based on my papers. I was invited to give lectures and talks all over the country; I was asked to serve on both the national and international monsoon panels that were a part of Charney's Global Weather Experiment. I was even offered jobs. Antonio Moura was the first to offer me a job in Brazil.

But the study of monsoon depressions was not where my true interest lay. These depressions could be predicted only a few days in advance. Surely a warning that one of these big storms was on the way was beneficial to society, but it wasn't going to change anyone's economic status or ensure enough food in times of hardship. I had my eyes set on the bigger, perhaps impossible prize: seasonal prediction.

CLIMATE, THE MOTHER OF WEATHER

The difference between the formation and dissipation of a monsoon depression and a monsoon drought is an example of the difference between short-term versus long-term forecasting—and also of weather versus climate.

"Climate is what you expect, and weather is what you get." This quote, often attributed (without any credible evidence) to Mark Twain, perfectly illustrates the fundamental difference between climate and weather, two terms we often use interchangeably.

While the actual weather can change from minute to minute, there is an expected weather for every minute of the year in every location on the globe. People living in Alaska are already used to the *climate*, the expectation that January will be very cold. It is the daily variation in this coldness—temperatures that are higher or lower than what is expected—that they experience as *weather*.

Why is it important to understand the difference between weather and climate? Because the physical and dynamical processes that are responsible for day-to-day weather are entirely different from those that determine climate and changes in the climate over time. Likewise, factors that determine how far into the future we can predict the weather are different from those that determine how far into the future we can predict the climate. For example, meteorologists have far more confidence in climate predictions one hundred years in the future than we do in the predictions of weather ten days from now. This is because, unlike future weather, which is strongly determined by the weather today, climate is determined by less erratic external factors.

Another term for *expected weather* is *mean climate*. The standard method to calculate mean climate at any point on Earth for any hour of the day is to take an average of all the weather data at that point for that hour for the previous thirty years. That is why you often read that climate is the average of weather.

What determines our mean climate? We've already gone over the biggest factors: the energy received from the sun, the energy emitted from Earth to space, the tilt of our planet away from the sun or toward it, and the combination of gases in our atmosphere. There are a few more important factors, including the planet's rate of rotation (which affects wind speed and direction), the mass of the planet (too small, and all gases would escape to space; too great, and gravity would preclude all weather), and the radius of the planet (which determines the speed, intensity, and size of storms). Finally geographic features—tall mountains, dry deserts, and wide oceans—also play a role in creating the mean climate.

While weather fluctuates constantly, the mean climate is relatively stable. Because the Earth is a sphere, the equatorial regions receive far more energy from the sun than they lose to space. This has been going on for millions of years, yet the tropical regions don't just keep getting hotter and hotter. Likewise, the polar regions receive far less energy from the sun than they lose to space, yet they don't get colder and colder. Why? Because both the atmosphere and the ocean are benevolent distributionists, continuously moving energy from the hot tropical regions toward the cold polar regions.

For example, the Gulf Stream and similar ocean currents push warm water from equatorial regions toward the north and south; similarly, hurricanes, typhoons, and other storms that originate in the tropical region transport heat north and south. Thus the atmosphere and the oceans act like Robin Hood, robbing excess energy from tropical regions and distributing it toward the cold polar regions that have an energy deficit, keeping our planet habitable by keeping our climate stable.

Stability is what we expect, but weather is what we get. What causes all that variation in the weather? The laws of physics require that in a rotating atmosphere like ours, where tropical regions are hot and polar regions are cold, the wind mostly blows from west to east, and the speed of the wind increases as we go higher up in the atmosphere. This is what happens on Earth and that is why we experience a strong jet stream in our upper atmosphere.

In 1946, Charney discovered that the wind shear can cause storms and other disturbances to generate and intensify by themselves, thanks to a process called "baroclinic instability," in which potential energy in the atmosphere is converted into kinetic energy. In other words, Charney gave a scientific explanation for why there is weather. This means that while climate is determined by cosmological factors (the ones listed above), the weather is determined by the properties of the mean climate, which is why sometimes we say that climate is the mother of weather.

Jule Charney was both a meteorologist and a climate scientist. A meteorologist observes and explains atmospheric processes and phenomena and predicts day-to-day weather. A climate scientist observes and explains the whole climate system—the atmosphere, biosphere, cryosphere, and oceans—and makes predictions and projections of future climate. Therefore, a climate scientist is a meteorologist, but a meteorologist is not necessarily a climate scientist. Both disciplines require an extensive understanding of math, the laws of physics, and computer modeling; however, society has assigned its own meanings to these terms.

The distinction was brought home to me when I went from MIT to Princeton. When I was at MIT, I told people that I was a doctoral candidate in the meteorology department, and this was generally met with a chuckle; in the 1970s, the public was not impressed with the accuracy of weather forecasts or the profession as a whole. But at Princeton University, the academic program concerned with weather and climate had a much fancier name: the Atmospheric and Oceanic Sciences in collaboration with the Geophysical Fluid Dynamics Laboratory of NOAA. Now when I told people what program I was in, their eyes widened with admiration—never mind that some of the greatest experts of geophysical fluid dynamics, like Charney and Lorenz, staffed the meteorology department of MIT and that the two programs had much in common.

Eight

My next indication that seasonal predictability might be feasible came just months after I defended my thesis. In the spring of 1976, I was back at the Geophysical Fluid Dynamics Lab in Princeton working as a postdoc in Dr. Manabe's group. On an otherwise ordinary afternoon, a friend of mine, Doug Hahn, a staff scientist at the lab, was walking down the corridor stopping everyone he passed to show them a sheet of paper—a plot of the year-to-year fluctuation in snow cover over Eurasia from 1967 to 1975. Doug was very proud of this figure because this was the first time ever that the area of snow cover had been derived from satellite data.

I knew that as far back as 1896, British scientists had used snowfall over the Himalayas to predict monsoon rainfall over India. (The idea was sound, but there wasn't much data to test the validity of it.) Intrigued by the new satellite-derived snow data, I stopped to study his graph, which immediately struck me as very familiar.

It turns out I had been staring at monsoon-rainfall data so long, the shapes and contours of its year-to-year fluctuations had been burned into my brain. Tracing my finger over the graph, I told Doug that the peaks and valleys of this curve looked similar to—and almost coincided with—the droughts and floods over India. At first glance, it looked like winter seasons of excessive snow cover over Eurasia were followed by summer seasons of monsoon droughts; seasons of deficient snow cover were followed by monsoon floods. "If we flipped this," I told my friend, turning the paper upside down, "it would look like a time series of the Indian monsoon rainfall."

Doug and I decided to plot both sets of data on one graph, and I proposed that we reverse the scale for the monsoon rainfall. It was amazing just how close the two curves were, suggesting that snow cover in a distant land—just like sea-surface temperature in a distant ocean—appeared to have a profound effect on the Indian monsoon. Our short paper turned out to be one of the very first published articles that showed a predictable relationship between snow cover and monsoons. It also gave a boost to the cryospheric scientists working on the role of snow and ice in climate.

I was starting to believe that in the case of seasonal climate, it wasn't initial conditions and those unpredictable butterflies that mattered most but rather boundary conditions at the Earth's surface. Just like it sounds, the term *boundary condition* refers to what happens at the edge of the atmosphere, whether that's land or sea. Snow cover is a boundary condition, as is sea-surface temperature. In a weather-prediction model, boundary conditions are not predicted but entered by the user (because it is the initial conditions of today that determine tomorrow's weather). Was I seeing a hint of order and predictability in an otherwise chaotic system? I needed more proof.

A few months later, I was back at MIT for a visit when Charney appeared in the doorway of my office. He had been invited to a conference in India, and, having never visited the country, he was intent on going, but he did not have a presentation topic.

"What do Indian people want to know?" he asked.

"Indian people want to know one thing, Dr. Charney," I said, pushing my chair back from my desk. "They want to know how to predict monsoon rainfall."

"And what do we have on that?" he asked, knowing full well this question was a good way to get me to start talking and never shut up.

He already knew about my Arabian Sea experiments, and now I told him about the apparent connection between the monsoon and Eurasian snow cover. "Dr. Charney, I'm starting to see that boundary conditions appear to influence monsoon rainfall," I said.

At this, my former adviser's eyes widened, and he began running his hands through his thick mop of hair. "Hold on," he said, and ran across the hall to his office. A few seconds later, he lurched back into my office, a paper in hand. I knew just which paper it was.

During his sabbatical in Israel, Charney had decided to try to come up with a dynamical theory of deserts, specifically whether human activity could lead to desertification.

Nearly all of the Earth's deserts—the Great Sandy, the Sonoran, the Sahara—are found between twenty and thirty degrees latitude. The reason for this has to do with the Hadley cell, a large-scale atmospheric circulation cell in which air rises at the equator, producing the wet climate of equatorial rainforests, flows poleward, and sinks around thirty degrees latitude. The sinking cool, dry air prevents the formation of clouds and rain. A cloudless sky means even more heat from the sun and an inhospitable climate for many plants and animals.

This is how nature forms and maintains a desert. But it occurred to Charney that there seemed to be anthropological ways to perpetuate a desert. Specifically, he suggested that the drought and expansion of Africa's Sahara Desert might have been accelerated in part by farmers overgrazing their animals in the dry grasslands at the edge of the region. This grazing, he posited, had stripped the land of vegetation and increased its albedo—its reflectivity—allowing more energy to escape to space and cool the atmosphere and thus further reducing the formation of clouds and rain. (This is somewhat similar to the vicious cycle that caused the Dust Bowl in North America.) Charney gave this theory the impressive moniker *biogeophysical feedback*.

The NASA climate model bore out Charney's theory. When he significantly increased the albedo of the Sahel region, much the way I had with the Arabian Sea, precipitation decreased, and the desert claimed more land. But ever cognizant of Lorenz's butterflies, my former adviser wanted to be sure chaos hadn't somehow infiltrated his experiment. So he ran the model three more times for forty-five days each, tweaking the initial conditions ever so slightly. In the end, all three simulations showed roughly the same results, demonstrating—as my experiments had—the influence of boundary conditions on rainfall, an influence that apparently eclipsed that of changes in the initial conditions.

These days, claiming statistical significance based on three model runs for forty-five days would be laughable (indeed, later, with better models and longer runs, this conclusion would turn out to be controversial), but Charney, the great fluid dynamics expert, could get away with that. But my results on the influence of ocean temperature and snow cover and Charney's results showing three similar averages of July rainfall were enough for Charney to accept my conjecture that boundary conditions might be crucial to predict Indian monsoon rainfall. We combined the two results, and Charney's lecture for India—which we titled "Predictability of Monsoons"—was all but ready.

A few months later I got a call from Charney, who was back in Massachusetts after his trip to India.

"The lecture went big," he told me. "Big."

Of course it did, I wanted to say. No Indian crowd had ever heard that you could use a modern climate model to predict the monsoon. Furthermore, Jule Charney was still the most recognizable name in meteorology, and he was saying that, despite all the butterflies in the world, seasonal prediction might be on the horizon.

There was one other thing about Charney's albedo paper that caught my attention, but it would take a few years for me to circle back around to it. At the moment, it was time for me to find a job.

After I defended my thesis, I was quite pleased by the number of offers for postdocs I received. Joseph Smagorinsky, the head of the lab at Princeton, even pulled some strings to get me a job offer from the World Meteorological Organization in Geneva. For a few weeks, I thought I was moving to Switzerland, but then Dr. Charney called and said he would like to meet with me.

He took me to a nice lunch that day, my first indication that he wanted to talk about something serious. When I mentioned the job offer from WMO, he told me the idea was, frankly, stupid. "You are a research scientist," he said, "not a bureaucrat." Finally, he showed his hand: He wanted me to take a joint NASA/MIT faculty appointment. It was an offer any scientist in our field would have killed for. I was so overwhelmed. I had to take some deep breaths and a few cooling sips of water. There was a part of me that still felt like a village boy, like the young man who showed up to his first big job in pajamas. Maybe I always would. But to get a vote of confidence from Jule Charney and an institution as esteemed as MIT was far beyond what I had ever thought possible for myself.

This position meant a lot of commuting between Cambridge and New York and then between Cambridge and suburban Washington, DC, to the Earth Science division at NASA's Goddard Space Flight Center in Maryland, where scientists observed and studied the planet's climate from space using satellites and spacecraft instruments. (Back then, I hardly thought of all the carbon the planes I rode every week were belching into the atmosphere.) At NASA, I sat next to some of the smartest people in my field and had access to one of its most powerful supercomputers.

In 1979, soon after I joined NASA, I invited Antonio Moura, a former classmate from MIT who was living in Brazil, to come visit. During his trip, we caught up with each other's scholarship, and Antonio told me he was working on untangling the dynamics of the severe and semi-regular droughts that occurred in northeast Brazil. Listening to him talk about the suffering these droughts caused the country's farmers, I began to think about the Indian monsoon and how the sea-surface temperature of the Arabian Sea seemed to have something to do with it.

A few days later, on a flight to Boston to visit Dr. Charney, Antonio and I started to look at some data he had stashed in his carry-on: the twenty-five-year time series of monthly mean rainfall over two weather stations in northeast Brazil.

"Let's try something," I said. "You tell me the rainfall for the year, and I'll guess if the Atlantic was warm or cool."

A smile spread across Antonio's face as he straightened the stack of papers on the tray table in front of him.

By the time we landed, both of us were sure there was a correlation between the temperature of the tropical Atlantic Ocean and rainfall over northeast Brazil. When we got back to Washington, a simple theoretical model that Moura developed during his visit and NASA's climate model both confirmed our hunch. When the tropical North Atlantic was cool, northeast Brazil was rainy. When the ocean was warm, the region was gripped by drought. We could not believe how beautifully a dipole pattern emerged. Our paper published in 1981 "went big," as Charney once said, in Brazil.

Score another for seasonal prediction.

My boss at NASA was Milton Halem. Milt had a broad forehead, thick waves in his yellow-gray hair, and more enthusiasm for earth science than anyone I had ever met. Charney had introduced me to Milt, and it was Milt who eventually brought me on full-time at Goddard and offered me one of the highest civilian positions available even though I was not a US citizen.

I'll never forget the look on Milt's face the first time I uttered the words *numerical climate prediction*. In fact, I had never before said it out loud to anyone. By 1980, numerical weather prediction had become the dominant way of making forecasts. But the idea of numerical *climate* prediction (employing the laws of physics and a computer model to predict seasons ahead) was still a veritable fantasy—and for most people, a laughable one. However, when I said those words to Milt, a visionary who was known for thinking big, his face brightened with excitement. He indicated he would love to see that scientific development happen right there in his lab at Goddard.

After my stint in India for MONEX, I returned to DC, intent on finally investigating the thing that had most piqued my interest about Charney's albedo paper. While my former adviser had focused on albedo, he had somehow failed to notice that the most significant change to the simulated rainfall was actually due to soil wetness, which he had only casually mentioned in additional experiments. When I pointed it out, Charney made several revisions in his manuscript and included an expanded discussion on the effects of changes in soil wetness, but this discovery stayed with me and made me wonder about the possible role of land (another boundary condition) in climate variability and predictability.

To find out, in 1980, I teamed up with Yale Mintz, an atmospheric scientist visiting NASA who had come up alongside Charney at UCLA. Together, we ran the NASA model using two wildly different scenarios. In one, we wanted the soil to be as dry as possible, so we essentially turned all the land in the world into a parking lot. No trees, no grass, no mud—nothing that could hold water that might evaporate into the atmosphere. In the other simulation, we made the land as wet as possible, transforming the whole world into a swamp, providing the maximum evaporation possible. In both models, we left the oceans alone.

What we found was shocking. When the whole world was a parking lot, continents were as much as thirty degrees hotter than they were when the world was a swamp; in Siberia, temperatures soared by forty degrees, as a cloudless sky permitted excessive heating from the sun. In the parking-lot model, precipitation over land fell by nearly 50 percent.

Although we had imposed extreme conditions that were impossible in nature, the experiment landed in the climate community like a grenade. For centuries, the common wisdom was that water that evaporated from the ocean was the source of rainfall over land. But what we had shown was that, since land-surface conditions and evaporation from land accounted for a stunning 65 percent of annual average rainfall, land was an essential component of the global hydrological cycle. As it turned out, the land didn't passively receive the weather; it actively created it. It was an almost heretical idea.

By the time our paper was published in *Science* (the top journal in our field) in 1981, I was convinced that it was possible, despite the butterfly effect, to make seasonal predictions based on ocean- and land-boundary conditions—and I was ready to say it publicly.

Back in 1968, I had stood behind a podium in Tokyo and contradicted Jule Charney, the most respected meteorologist in the world. In 1981, when I arrived at the European Centre for Medium-Range Weather Forecasts in Reading, England, I was preparing to contradict the man who had taken Charney's place as the god of all things weather, Edward Lorenz. At least, I thought to myself as I checked into my hotel, back in Japan I hadn't known enough to be nervous. Now I knew full well what I was doing: standing up in front of the revered scholar who said long-range forecasting was not possible and announcing that, in fact, forecasting seasonal mean was possible.

Edward Lorenz was not a cruel man. But he was direct. And he suffered no fools. Recently an old friend of mine from MIT confessed that he had left the field after graduation because of a single sentence that Lorenz had said to him after his thesis defense. A pillar of brilliance, Lorenz seemed to possess a preternatural ability to suss out the shaky spot in an argument, the weakness in a job candidate, the one tiny detail in your experiment you hadn't thought of. Lord have mercy on the man who tried to contradict Edward Lorenz.

Of course, I was not contradicting Lorenz, I was merely pointing out, perhaps for the first time, that in certain situations—when tropical oceans were unusually warm or cold or when land surfaces were unusually dry or wet—there were exceptions to the butterfly effect.

In Reading, I would be delivering a lecture with two parts based on two research papers. The first, Lorenz himself had encouraged and assisted me with; it was about the difference in predictability between planetary waves—atmospheric currents that circle the entire globe, like the jet stream—and synoptic waves, smaller atmospheric systems, such as cyclones. My work showed that planetary waves had much higher energy and longer predictability than synoptic waves, and therefore monthly and seasonal averages that were dominated by planetary waves might be predictable. The theoretical work on weather predictability and the limits of weather prediction were defined by the rate at which the initial errors grew, but there was no such metric to define the limits of predictability of monthly and seasonal averages. My paper proposed to define the predictability of monthly and seasonal averages as the ratio of signal due to boundary conditions and noise due to small random changes in the initial conditions (the butterflies).

The second paper I thought of as the punch line of the first. It contained the results of the various numerical experiments concerning boundary conditions that colleagues and I had done over the past five years, the ones on sea-surface temperature, snow cover, and soil wetness. I had not discussed these ideas or the results of these experiments with Lorenz, and I was nervous about how he would react to my lecture. I don't generally experience stage fright before big talks, but on that gray September day in 1981, I was stressed.

As soon as I took the stage in the ECMWF seminar room, I spotted Lorenz among the two hundred or so scientists seated before me; he was tucked into the fourth or fifth row, his thinning hair pushed neatly to the side. Lorenz was, by a large margin, the most prominent scholar at the conference, and just like Charney all those years ago in Tokyo, he seemed magnetic, as if he drew all the energy in the room around him. Swallowing hard, I lowered the microphone and began my talk.

In it, I argued that in this world, there were some forces so big and commanding—long waves and boundary conditions among them—that not even the chaos of initial conditions could change their influence.

I showed that boundary-forced effects can be so large that not even a billion butterflies could make them unpredictable. Finally, I confidently stated that there was a scientific basis for numerical seasonal prediction in the midst of chaos.

At the end of my talk, I asked if there were any questions. A few attendees raised their hands, but not Lorenz. I felt my shoulders start to relax. But after the event ended, as people were milling about and shuffling out of the room to have coffee, I saw Lorenz emerge from the crowd—and head my way. I steeled myself.

"Uhh." My former mentor made a sound like he was thinking, trying to find just the right words to describe my presentation. Then, in his trademark low and slow intonation, he said, "I didn't know you were working on these things. That was very interesting." It was gratifying that when Lorenz submitted his manuscript, he referred to one of my papers regarding the role of slowly varying ocean temperature.

Modest praise from anyone else, but from Lorenz, this was nothing short of a benediction. And as the two of us left the center that day on our way to get coffee, I thought about the road ahead. During my half a decade in the United States, I had made some major discoveries and formulated a hypothesis for the feasibility of numerical seasonal prediction. My desire to make an impact on society and improve the lives of people like those I had grown up with now rested on a firm scientific foundation. But enacting real-world change would require challenging the status quo of an entire academic field as well as the entrenched practice of using statistical methods for seasonal forecasting. In many ways, the work was just beginning.

This became abundantly evident to me when I was invited to take part in a discussion with scientists at the World Meteorological Organization in Geneva who were in the process of organizing a major international conference on climate. I was so enthusiastic about the idea of numerical seasonal prediction that, naturally, I proposed that the conference should be called Climate Prediction.

I was quickly told that it would be impossible to convince the international community that science had advanced enough to have a conference on climate prediction. But I kept at it, arguing passionately that the title of the conference should at least contain those words. "How about the Physical Basis for Climate Prediction?" I asked. Finally, the powers that be at the WMO budged and the 1982 international conference on Physical Basis for Climate Prediction was held in Leningrad, Russia.

BETTER TOOLS, BETTER FORECASTS

My hypothesis of seasonal prediction got another boost from the pioneering satellite data Doug Hahn had been showing off in the corridors of Princeton. Back then, in the mid-1970s, climate scientists everywhere were having similar experiences. Confronted with new and at times confounding information from outer space, my colleagues and I had to reconsider old ideas and incomplete hypotheses.

Until the 1970s, the perception of the public was that weather forecasts were more often wrong than right. While this was partly due to people's tendency to remember only inaccurate weather forecasts, that perception did begin to change in the 1980s. Today, most people would agree that weather forecasts have vastly improved. If you were to ask someone who is not a trained meteorologist why that is, he would probably think of the many, many images of white, swirling weather systems taken from space and mention the advent of weather satellites.

But a satellite image, no matter how sharp, can tell us only what the weather is right this moment, not what it will be in the future. As we know, a weather forecast is created by complex mathematical models consisting of billions of equations solved by a supercomputer. These equations require an accurate description of the initial conditions of the atmosphere, but how those initial conditions will lead to future weather cannot be gleaned from a picture.

In fact, initial conditions themselves cannot even be gleaned directly from the actual *data* provided by satellites, which measure only the radiation leaving the Earth. Therefore, over the past fifty years, scientists have developed complex techniques to calculate temperature from that radiance, a method called retrieval of satellite data.

Once the temperature is derived, it is combined with conventional data from ground stations, ships, and aircraft to produce the most accurate initial conditions possible. This is called data assimilation, the output of which can then be fed into the supercomputer to generate a forecast. The assimilation process is a colossal task. The computer codes to create accurate and consistent initial conditions are longer than the computer codes for making actual weather predictions, which themselves can be as large as a million lines of computer codes.

It took some time for scientists like Jule Charney and others to show that this process would indeed work to produce better forecasts, but eventually the evidence was clear, especially for the Southern Hemisphere, which does not have a sufficient density of ground stations.

In addition to these developments, faster computers have played a major role in improving weather forecasts for two distinct reasons. First, the faster the computers, the less time it takes to prepare the initial conditions and start running the prediction model. Forecasts are made for each day in time steps of a few minutes, and at each time step, billions of equations are solved until the end of each day. Therefore, extremely fast computers are needed to ensure that it takes less than thirty minutes to make a forecast for the next twenty-four hours.

Second, the faster the computer, the higher the resolution of the prediction model, the better the description of things like storms and hurricanes. Therefore, the models can be more accurate. Ironically, as our models reach higher resolution, initial errors also grow faster, so the limit of skillful weather prediction becomes a competition between how small the initial errors are and how fast these errors grow, thanks to the butterflies of chaos. We continue to improve our forecasts with more accurate initial conditions.

About fifty years ago, the spatial resolution of weather-forecast models was around one hundred fifty to two hundred kilometers. Today's weather-forecast models have a spatial resolution of ten to twenty-five kilometers. This is a great thing! To make forecasts, we have to solve the differences between the weather at two different points in space and time. To minimize uncertainty, we want the points to be as close as possible. The greater the resolution, however, the more calculations and computer power you need. A few centers in the world are now developing computational and technological infrastructure to run one-kilometer global models for weather and climate predictions and climate projections. This will require computers that are a *trillion* times faster than the computers of the 1970s.

These advances have been made possible by advances in computer technology and computer-chip manufacturing. During the past half century, there has been a billion-fold increase in the number of transistors on a chip, meaning computers used at the leading weather-prediction centers in the 2020s are more than a *billion* times faster than computers used in the 1970s. Powerful supercomputers consume lots of electricity, and the cost of electricity has become a serious obstacle in acquiring faster computers. Several years ago, the city of Boulder, Colorado, informed the National Center for Atmospheric Research, one of the leading supercomputer centers for weather and climate research, that the city could no longer provide sufficient electricity for the center's next-generation computer. NCAR had to move its supercomputer to Wyoming, which could provide sufficient electricity at a reasonable cost.

Computer companies and research scientists are also trying to develop artificial intelligence and machine-learning techniques—which do not need supercomputers to solve billions of equations—to produce short-term weather forecasts at extremely high spatial resolution. Some preliminary results are sufficiently encouraging to be of real concern to the supercomputer and dynamics and physics-based weather-forecasting centers, although AI and machine-learning techniques will likely never replace the extremely high-resolution, complex physics-based models because they need the past data from such complex models to train the AI models.

Nine

I think of my years at MIT, Princeton, and NASA as charmed. Professionally, I could not have been luckier; I was surrounded by brilliant colleagues. Scientifically, I was on a roll. But during this time of victories and success, my personal life seemed to be mired in chaos, roiled by events—both good and bad, sweet and bittersweet—completely outside of my control.

For one thing, I had become a father.

I had finally met Premda, my wife, while I was working at the institute in Pune. Just before going abroad to the United States and Tokyo, I traveled to Ballia City, where she was staying with her parents. (By tradition, she could not come to my house in the village for five years, when final marriage ceremonies would take place.) It was a short and awkward meeting; I was twenty-one, she was eighteen, and we had nothing to talk about. During our long wedding ceremony, I had squinted through the gauzy curtain that separated us and wondered if my bride was feeling the same combination of reticence and resentment as I was. I might have seemed like a catch to her father—I had one of the best exam scores in the district and the promise of steady government employment ahead of me—but who knows what Premda thought. Like most Indian women obligated to enter into arranged marriages, she was not asked about her opinion, and that day in Ballia City, she didn't share it with me.

A few months later, in Tokyo, I bought a Japanese pearl necklace for her. Walking back to my dorm, I was filled with excitement and anticipation of family life for the first time. Sadly, our next meeting didn't go much better.

A year or so later, I confided to my roommate Venkataramanaiah Krishnamurthy (the same friend who would proofread my thesis) that the next time I returned to the United States from India, I would have my wife with me. Krishnamurthy strongly encouraged me to do so.

Alas, the next visit was no different from the earlier ones. We were both young and immature.

Despite the lack of affection or admiration between the two of us, my mother insisted the marriage would work, and she began inviting Premda to the village when I came home from Pune and pushing us to spend time together. The last time I had seen my wife was the week before I left for MIT.

Then, in early 1972, a letter from my brother Kanhaiya arrived in Cambridge. *I have good news and bad news,* he wrote. The good news: I was a father. The bad news turned out to be classic Indian ribbing: the babies were twin girls (each of whom would one day require a large wedding dowry). Sitting in my apartment in Massachusetts, the radiator banging and hissing beside me, I read the letter over and over. For so long, the things that had happened in my personal life felt so random—my father's death, my marriage. And now daughters? A few weeks later, I received another letter saying that one of the twins had died. None of it seemed real.

Today it feels preposterous to think that I didn't meet my daughter for the first two and a half years of her life. That I never got to hold her when she was a baby or see her take her first steps. But all that distance seemed so normal to me at the time; my model of fatherhood was one of both emotional and physical separation. My own father was rarely at home with his two wives, and he almost never slept in the house. In our village it was considered very strange for a man to live too closely with his wife, just as it was normal for a groom not to learn his bride's name until after the wedding. Most of the men in my village who had jobs left their wives and families for long stretches to work in big cities and make money, just as I was now.

In 1974, I finally had the chance to go home and meet Pooja, a spunky toddler who wasn't too interested in visiting with me, given that many cousins and second cousins were vying for her attention there in Mirdha. That was okay; I was content to just watch her laugh and play. I decided right then that I would make it a priority to visit Mirdha every year, no matter what. Soon I would be back in India for good, and when I was, I would make sure my girl attended all the best schools, just like my father had for me.

But then suddenly, the India I loved and longed for began to change. In the summer of 1975, Prime Minister Indira Gandhi, facing fierce opposition to her rule, declared a state of emergency, effectively transforming India from a democracy to a dictatorship overnight. For the next twenty-one months, my home country became an authoritarian state where constitutional rights were nonexistent and critics of the regime were rounded up and thrown in jail without so much as ten minutes in front of a judge. The press was censored. The poor were forcibly sterilized. At MIT and then Princeton, I watched from afar, scared for my family and worried about the future of India.

A year after the emergency began, I received a letter from my mother with news about my brother. In truth I had been expecting a letter like this for some time. My older brother, the would-be wrestler, was very active in politics, and while he supported Indira Gandhi's regime, India was such a chaotic place, it didn't seem unlikely that one day he would end up in jail. But my mother's letter conveyed news about a different brother: it was Kanhaiya, my younger brother, who had been arrested and detained in the district jail. Kanhaiya wasn't political at all; at his college, he was a sports champion and a natural leader, handsome, athletic, well liked, and followed by a crowd every time he went to a tea stall. Apparently, he was so popular that a police officer in the district got suspicious he was an organizer, and the hunch of that officer was all it took for my brother to wind up in a jail cell in Ballia City.

My mother was naive to think that just because I had an American address, I could do something about the situation, but I must have been full-on delusional because I booked a flight to India immediately. Perhaps I had a little of my father in me, the conviction that any obstacle could be overcome by simply meeting officials and pleading your case.

As it turns out, the fact that I was from the United States did at least get me in the office of the superintendent of police in Ballia City. The short, middle-aged man studied me with apparent fascination as I entered the windowless room, where he had his feet propped up on the desk. Staring past the soles of his boots, I tried to tell him that my brother was not an activist, a troublemaker, or a rabble-rouser, just a popular young man with a large social circle. The superintendent said nothing, just sucked on a bidi, a pungent cigarette wrapped in tobacco leaves. Smoke clung to the low ceiling. Outside the door, two peons sat on stools awaiting orders.

I was beginning to think this was a bad idea. In the India of my childhood, my plea might have been met with reason. But in the India under the national emergency, this police superintendent had sweeping, unchecked power. He seemed to almost enjoy watching me beg. Finally he spoke.

"I have a solution," he said, and the tightness in my chest subsided just slightly. "How about I take him out—and put you in?"

To walk out of a room in India without so much as a handshake or a word of gratitude is a profound insult. I did it anyway. I went home and told my mother that I was sorry, but there was nothing I could do. In fact, my presence there might have made the situation worse. As she had when my father took me to boarding school, she had stopped eating, too concerned about her son to entertain the thought of food.

On that visit, just like all the others, I spent time with Pooja, my little girl, who was growing like a rice seedling in monsoon rains. Watching her amble around the grass with her cousins, her wispy black hair moving in the hot wind, I was filled with joy—and worry. I wanted a good life for her, one that wouldn't be decided by monsoon droughts or political dysfunction. But like my mother, I felt helpless, unable to make that happen.

Days later, as my plane taxied down a runway in Delhi, I wondered what my future held. Every other time I left India, I had believed it was temporary. But on this trip, things felt different. How could I make any kind of a change in a place where the newspapers weren't allowed to print the truth and people were thrown in jail for nothing? I had the ear of powerful people in the United States, had access to some of its most esteemed institutions. Maybe, I thought for the first time, I could do more if I stayed in America.

About two years later, the butterflies flapped their wings again, delivering something completely unexpected into my life. She had brown hair and a subtle Wyoming accent. She was a Greek American; her name was Anastasia, Anne for short; and she was the most beautiful person in Cambridge, Massachusetts.

By the time I returned to MIT for the dual appointment, I had known Anne for about five years; her husband was in the meteorology department. While he'd earned his Ph.D., she'd earned the money, working as an assistant in the university's Architecture Machine Group, a think tank dedicated to the interactions between humans and ever-smarter technology headed by Nicholas Negroponte. (The Greek American architect would go on to create MIT's Media Lab with Jerome Wiesner, former science adviser to JFK and the man who suggested Charney's Global Weather Experiment to the president.) I saw Anne frequently at department get-togethers and parties and noticed how kind she was, how she always seemed to be laughing. After her husband got his Ph.D., they moved to California.

So when I learned that Anne and her husband were getting divorced, I was quite sad for them. I even tried to recruit a group of our friends to help keep them together. That made perfect sense to an Indian boy like me, someone who understood marriages to be carefully negotiated, practical arrangements. "You are naive," my friends told me. "It doesn't work like that."

I returned to Cambridge after my postdoc year, and Anne, who had stayed with her parents in Denver for a few months, also came back. Before, she and I had hung out in large groups among friends, but now the two of us sometimes went out to dinner alone. On these occasions, I encouraged Anne to reconcile with her ex-husband, completely oblivious to the feelings developing and deepening between us.

Finally she told me, "I care so much more about you than I ever did about him." Her words hit me like a rock to the forehead. Things moved fast then; my life changed entirely. Soon I learned other powerful words.

I love you.

No one had ever said those words to me or even around me. Where I came from, love wasn't so much a word as an action—a warm meal, a prayer, a ghee candle lit in your honor. But I liked the word; I liked the way it made me feel. And I liked that lovely, intelligent Anne was willing to take a chance on me—a foreigner from a poor family with a whole complicated life behind him. Her presence, the way she listened so carefully, gave me a grounded feeling, a remedy to the turbulence that seemed to consume my personal life.

Until then, I'd felt firm in the idea that I would never get married again, no matter how much of a mismatch Premda and I were. But Anne's words made me think of a different future.

When I told Anne that I would always send a portion of my income to support my mother and Premda and pay for the education of my younger brothers and Pooja (who would attend the finest boarding school in Varanasi), her reaction was simple: As long as we have enough money to eat and live, you can send all the money you'd like. Choosing to be with Anne was significant because it was a choice, my choice, and I made it with confidence.

When I told my mother about Anne, she was ecstatic. She had been waiting for me to marry again for so long that she did not even care that I was marrying a non-Brahmin from America. The idea of her son living without the care of a wife was anathema to my mother, who spent so much of her own life devoted to the well-being of her family.

I had to burst her bubble and tell her that before I married Anne, I would be divorcing Premda. My mother, a second wife herself, was confused—what was the problem with having two wives?

After I convinced my mother that I needed to do things the official and the legal American way, my mother, Premda's parents, Premda, and I hammered out a separation agreement that ended the arranged marriage. It was not something that I took lightly. In fact, I had spent years vacillating between trying to convince Premda to come to the United States and thinking I should leave well enough alone; I had lain awake many nights trying to figure out how to make our relationship work. But it turned out that Premda and I were like a nonlinear equation: two variables that could not be reconciled and that were affected by a whole host of other variables—her family, mine, her wishes, mine. Falling in love with Anne made me realize I would never solve the equation with Premda and that it was time to move on, to embark on a marriage based on love.

On an early autumn day in 1979, I spent the morning working at NASA and then met Anne at the Montgomery County, Maryland, courthouse. I wore a gray suit and Anne wore a flowing red dress. We invited only my boss, Milt Halem, his wife, and two of our closest friends. On the day I married Anne, I knew I was also committing to something else—I would stay in America for good.

Among the Indian graduate students that I had known at MIT, I was quite notorious for insisting that the values of the West would not influence my thinking or behavior. I would never, for example, marry a divorced woman. I would not stay in the United States, and I certainly would never buy a home here.

But by 1981, around the time I was speaking in front of Edward Lorenz at the ECMWF, I had done all three of those things.

EL NIÑO Y LA NIÑA

I don't know if I would have become a climate scientist had I not been born in the monsoonal country of India, but working on seasonal prediction for the Indian monsoon did not mean that I had to stay in India. Climate science is a global undertaking, one that requires perhaps a greater amount of international cooperation than any other human endeavor. While even allied countries compete politically and economically, they tend to take a more collaborative stance when it comes to weather and climate prediction. After all, the storm that ravages one country today will arrive in another country tomorrow. As El Niño—one of the most well-known and widespread natural climate phenomena—shows us, effective climate science requires a global perspective.

The earliest recorded mention of El Niño in the context of climate appears in 1881 in the minutes of the annual meeting of the Peru Geographical Society, held in Lima. During that gathering, Peruvian sailor Camilo Carrillo reported that every four or five years or so, around Christmas, he and his fellow seamen had noticed a southward-flowing ocean current of warm water. They called it El Niño de Navidad, Spanish for "Christ Child."

Usually, the ocean current along the Pacific coasts of Chile, Peru, and southern Ecuador flows from south to north. The Peru Current, as it's known, typically brings cold water north, keeping those coastal areas cool and dry. However, during the years of El Niño, the ocean temperatures remain unusually warm for a season or longer, and the direction of the Peru Current reverses, bringing with it abundant rainfall. During these years, river basins swell with water and the Peruvian deserts blossom.

At the dawn of the twentieth century, long after the Peruvian fishermen had started noticing this band of warm water, the British government was hard at work trying to figure out another unusual phenomenon halfway around the world—monsoon rainfall. In 1904, Gilbert Walker, who had been appointed director general of the India

Meteorological Department and tasked with finding a way to predict monsoon droughts, did not have any information about the temperature of the Pacific Ocean or the experiences of Peruvian fishermen. What Walker did have was data on air pressure and rainfall taken from observations in many of the British colonies, so he calculated correlations between Indian monsoon rainfall and the air pressure over the stations for which data was available.

He discovered that when the air pressure was higher at the Australian weather stations of Darwin and Sydney, the air pressure was lower at the Tahiti and Buenos Aires stations, and vice versa. In other words, when the western Pacific and the Indian Ocean experienced high pressure, the central and eastern Pacific regions experienced the opposite.

Walker had no way of knowing that this large-scale pressure seesaw—which he named Southern Oscillation—was intimately related to El Niño. For years following Walker's discovery, meteorologists studied the pressure patterns and oceanographers studied ocean temperature patterns, but neither group recognized that they were studying two sides of a single, strongly coupled ocean-atmosphere system.

Finally, in 1969, Jakob Bjerknes published a groundbreaking paper suggesting that the two phenomena, the Southern Oscillation and El Niño, were connected. He had come to this conclusion after analyzing ocean-temperature data from 1957 and 1958 and noticing that during El Niño, warm ocean temperatures were not confined to the coastal regions but covered a larger region of the tropical Pacific Ocean than previously observed.

Bjerknes pointed out that for the long-term average climate of the Pacific region during non–El Niño years, the warm air rose and gave rain over the regions of warm ocean temperature and lower pressure in the western Pacific and colder air descended over the colder ocean region of the eastern Pacific. This circulation helped maintain the steady trade winds blowing from east to west near the equator. He named this the Walker circulation.

What Bjerknes had discovered was a mechanism for the interaction between wind circulation, ocean temperature, and rainfall that maintained the Walker circulation. He further suggested that during the El Niño years, when the ocean warmed and surface pressure dropped in the eastern Pacific, the trade winds got weaker and sometimes reversed, blowing from west to east, while the areas of heavy rainfall in the western Pacific shifted eastward to the central Pacific. This shift in the rainfall and the associated warming of the atmosphere by the latent heat of condensation is what produces the worldwide change in climate during El Niño.

The opposite phase of El Niño, when the eastern Pacific Ocean is colder than normal, the areas of high rainfall in the western Pacific shift westward, and the trade winds are stronger, is called, naturally, La Niña. Bjerknes could not explain how and why this turnaround occurred. Scientists' understanding of the dynamics of both the atmosphere and tropical oceans has greatly improved over the past sixty years, and several mechanisms have been proposed for transitions between El Niño and La Niña, but it remains an area of active research.

What is certain is that El Niño has and continues to cause chaos in the natural and man-made world. During the twentieth century, more than a quarter of all years were El Niño years and each had its own distinct environmental and economic impact. The El Niño of 1957/1958, for example, the one that helped Bjerknes discover the Walker circulation, wiped out vast expanses of California's kelp forests. The 1965/1966 El Niño devastated the guano market in Peru—an untold number of seabirds died from starvation, and heavy rains washed much of the guano off the rocks where it had been harvested for fertilizer for centuries. In 1980, my former MONEX colleague Dev Raj Sikka showed that, with some notable exceptions, major droughts over India occurred during the years of El Niño, the single most important predictor of monsoon rainfall, even today.

In 1982, atmosphere and ocean scientists gathered at the Geophysical Fluid Dynamics Lab in Princeton to plan a massive field experiment to observe, model, and, hopefully, predict El Niño. At this meeting, I proposed my theory that slowly varying sea-surface temperature changes could produce predictable changes in atmospheric circulation and rainfall, a hypothesis I hoped would encourage the improvement and application of coupled ocean-atmosphere models for seasonal prediction of rainfall.

Ironically, as if to highlight the need for such an effort, at that time, a powerful El Niño was brewing in the tropical Pacific Ocean, but the scientists had no clue; some insisted that an El Niño wasn't even possible that year. A few months later, Christmas Island seabirds abandoned their nestlings in droves to find food, and fur seals and sea lions starved to death by the millions.

Ten

It was at the Physical Basis for Climate Prediction conference in Leningrad where I, a newly married, now permanent American resident, met Gabriel Lau, a scientist working under Dr. Manabe in the Geophysical Fluid Dynamics Lab (GFDL) at Princeton. Lau was there to give a presentation on an experiment Manabe and he had run, one that was now famous because it was the world's first realistic simulation of global climate and especially the climate of the tropical atmosphere, which was notoriously difficult to simulate.

However, the model used climatological SST (sea-surface temperature), the monthly average for the past thirty years. That meant that each January, the model ran with the same sea-surface temperature value, and the same one each February, each March, and so on. But based on what I had found in my experiments, I knew that anomalies in sea-surface temperature could produce wildly different weather patterns and that the GFDL model would be even more realistic if it employed the *observed* SSTs—the actual values, not simply the climatological average. My paper at the same conference showed that year-to-year changes in ocean temperatures were the most important factors in explaining year-to-year climate variability and could be the basis for dynamical seasonal prediction.

After Lau was done with his presentation, I rushed over to ask why he and Manabe had not used ocean temperatures that changed from year to year. He said that such a global gridded data set did not exist. What he meant was that the raw observations, taken from ships and weather stations scattered about randomly, had not been arranged into tidy rows of latitudes and longitudes, and that was the gridded data that computer models needed. I thought that there should be a way, especially at Princeton's Geophysical Fluid Dynamics Lab, one of the world's foremost research institutions, to create such a data set. I suggested that he and Manabe should acquire the data and run the model again with year-to-year changes in sea-surface temperature. He politely informed me he would relay my message to my former adviser.

A few months and a few phone calls from me to Princeton later, Lau and Manabe still hadn't done what I was suggesting. I felt so passionate about this idea, so sure it could transform our long-term forecasting capabilities, that I couldn't sit in Maryland and wait any longer. As a NASA employee, I had access to one of the agency cars, a white sedan with the NASA logo painted on the sides. I had never utilized this company perk, never driven much at all in the United States, but I knew I needed to get to Princeton as quickly as possible and convince the scientists to acquire the observed sea-surface temperature data and run the model again.

It was a sunny weekday morning and the New Jersey Turnpike was filling up with commuters. Behind the wheel, I let my eyes glaze over a bit as I practiced my plea to Manabe. I might have vaguely noticed the blue and red lights swirling in my peripheral vision, but they didn't fully register. It was the siren that jolted me from my thoughts. With my already excited heart pounding even harder, I pulled the little sedan over to the shoulder, put it in park, opened the door, and began to step out to meet the officer, which is the custom in India.

"Sir, please step back into the vehicle!" came a gruff shout from behind me.

I did as I was told and slipped back into the seat. The sound of boots on gravel got closer, and then a stern-looking man stood at the driver's side window.

"Can you please read what is written on the dashboard?" he asked, a question that sounded more like a command.

Confused, I searched the console in front of me for something that I could read out loud. I had been in the car for a few hours now and hadn't noticed the sticker below the speedometer, although it was right in front of my face.

"'This vehicle limited to fifty-five miles per hour . . .'" I said, trailing off to present a contrite grimace to the policeman.

I decided not to tell him I had been breaking the rules—albeit unknowingly—for the sake of improving the most powerful climate model in the world.

When I finally arrived at Princeton, I learned why Manabe hadn't yet acted on my plea—he didn't have the gridded values of year-by-year ocean temperatures. It was the same answer Lau had given me in Leningrad. But, I argued, the National Oceanic and Atmospheric Administration had the raw data for sea-surface temperature, and although it would require tremendous staff time to analyze the past data and grid them for the model, doing so was absolutely essential. My work on boundary conditions had convinced me of that. By the end of the conversation, Manabe agreed to get the data and incorporate year-by-year monthly observed sea-surface temperature into the model.

It took a few months, but eventually Lau and Manabe ran the model using fifteen years of observed values of ocean temperatures. Now, instead of the same twelve numbers over and over again, the model used a different number for each month of fifteen years—the actual sea-surface temperature. It was immediately evident that the effort had paid off when the model began to produce more realistic simulations of the current climate and its variability.

In fact, this endeavor was so successful that a few years later, the

WMO launched the Atmospheric Model Intercomparison Project, urging all the global atmospheric models in the world to use the year-by-year values of observed sea-surface temperature. I felt tremendous relief when this happened; never again would I need to break the law in a government vehicle to fight for the integration of observed ocean-temperature values!

As the models began to change in response to a deepening understanding that the ocean and land play significant roles in climate, it was clear to me that our institutions needed to change to reflect this knowledge as well. I treasured every moment that I was privileged enough to work at NASA, but doing the multidisciplinary science of atmosphere-ocean-land interactions was becoming increasingly difficult due to the conventional way the agency siloed scientific disciplines and experts.

For instance, the Goddard Space Flight Center had three climate divisions: atmosphere, oceans, and hydrology (the science of water on land). Each division was headed up by a chief and cordoned off into three or four branches. This made collaborating with anyone who was in a different division or branch than yours quite difficult, fraught with paperwork and special permissions. NASA wasn't the only organization that looked this way; back in the mid-1980s, interdisciplinary work was not yet popular, and all the major research institutions were honeycombed. But I and a growing number of my colleagues were convinced that if we were going to develop a robust scientific basis for seasonal prediction, we needed to get the experts together in one room.

One possible option was to establish a research center at a university. When I spoke with program directors at NASA, the National Oceanic and Atmospheric Administration, and the National Science Foundation, I got a clear impression that they were all interested in supporting my novel research idea of exploring the feasibility of dynamical seasonal prediction. NOAA was interested because prediction was one of the central missions of the agency. The NSF was keen to support research that could advance science's understanding of predictability beyond weather. NASA was particularly intrigued because I was suggesting that its satellites could play a role in seasonal prediction (by observing boundary conditions such as global ocean temperature and soil wetness). For me, the idea of not working for the government while the current head of that government, President Reagan, was relentlessly criticizing and demeaning federal employees was an added attraction.

I reached out to scientists I had met at MIT and Princeton who had gone on to university appointments about setting up such a center at their institution. After I had several conversations with universities and even a few corporations, the University of Maryland was the front-runner; administrators said they could offer three tenured faculty positions for what we would call the Center for Ocean, Land, Atmosphere Interactions (COLA). It would be staffed by myself and two former MIT classmates, Mark Cane and Ed Sarachik, a postdoc under Jule Charney who was now working as a research scientist at Harvard. These were two very well-known and well-respected scientists who, I knew, would advance the science and immediately lend credibility to our new center.

In 1984, it was time for me to resign from NASA.

Leaving NASA was incredibly hard, especially because my boss, Milt, had become such an important person in my life. He had gotten me a high-ranking government job. He made sure I was able to have as much time with the supercomputer as I needed, and he'd given me a handful of staffers to help run my experiments. Milt and his wife were among the only guests that Anne and I invited to our tiny wedding. But Milt was supportive (if reluctantly at first) of my decision, knowing I had run out of runway at NASA.

As soon as I resigned my position, I received bad news from UMD. In a decision that foreshadowed the discord to come, UMD rescinded the offer of two of the three tenured positions, saying that they would have to be tenure-*track* assistant professorships, appointments that came with less money, less job security, and less prestige. Immediately, both of my old classmates from MIT refused the offer. I couldn't blame them.

Instead, we hired two brilliant postdocs from Princeton, Jim Kinter and Jim Carton. Jim Kinter, an atmospheric scientist, was brimming with confidence, and in the years to come, he would translate all my rambling harebrained ideas into convincing research proposals. Jim Carton was an expert in ocean science.

In addition to the two tenure-track faculty members, we hired other scientists: Ed Schneider and David Straus. Ed came from MIT, where he had been one of Charney's postdocs, and could write a model like no one else I knew. Everyone who knew Ed inevitably referred to him as "the most brilliant person I've ever met." The group received a huge boost of confidence when David—another Charney acolyte—resigned his high-ranking civil service position at NASA to join our fledgling organization.

Every one of these men could have taken a job in private industry and made far more money. They could have enjoyed tenured positions at esteemed universities. They could have gotten lifetime appointments to the world's finest research institutions. But each of them chose to come work at the center because they liked the scientific challenge and believed in the mission—advancing seasonal prediction for the benefit of society. The new center at the university had a good start, and the number of applications to the Ph.D. program at UMD quickly doubled.

With that, we got to work.

As I attempted to expand the paradigm of prediction beyond weather in the United States, the science and the practice of routine weather prediction fell farther and farther behind back in India. Politically, things were much improved. In 1977, Indira Gandhi had abruptly declared an end to the state of emergency and released all political detainees, my brother included. During the next elections, a new prime minister was voted in. There wasn't much progress in the world of Indian weather prediction, however; the country's meteorologists still relied on unreliable upper-air observations and outdated regional models. Thanks to my research on monsoon predictability and my frequent work visits, I had become friendly with Yash Pal, India's secretary of science and technology, whom I badgered constantly about the improvements the government needed to make to modernize its weather and climate enterprise and advance its forecasting capabilities.

For one thing, I told the secretary, the Indian upper-air observation stations were well known for producing wildly inaccurate data. While most countries purchased their observation instruments from scientific-instrument manufacturers, the India Meteorological Department insisted on making their own. Imagine if the postal service built their own delivery vehicles or the police fashioned their own weapons. The inadequacy of the system was quite apparent in the values it supplied to weather-forecast models around the world—nonsensical numbers with fantastical rates of change.

The other thing I told the secretary was that India needed to invest in a modern supercomputer and stop relying on ancient machines like the ones I had worked on in Pune and that were still in use in the mid-1980s. This, I knew, was a likely impossible task, since all the best supercomputers were built by American companies, and the American government would never authorize the sale of one to India, a Soviet ally at the height of the Cold War. An Indian supercomputer might become a Soviet supercomputer, the thinking went, and be put to use developing nuclear weapons or cracking American military code.

Still, I kept up my one-man campaign to modernize India's weather-forecasting system. Finally, in January 1985, during one of my annual trips to India when I was once again trying to persuade the secretary, he grew exasperated.

"I'm meeting with the prime minister soon. So just write it all down," he said, placing a pad of paper in front of me. "Write down exactly what India needs to do." He directed me to a room next to his office and asked for a handwritten report on numerical weather prediction, including how many people and what kind of computer would be needed to establish a NWP center in India. Even though I assumed this document would go where all my other advice seemed to go—nowhere—I wrote a short paper, emphasizing that the critical component was a supercomputer.

Sometime later that fall, my list, or at least one of the items on my list, apparently came up at a very important meeting between President Ronald Reagan and Prime Minister Rajiv Gandhi. The two had met for tea in New York at the Waldorf Astoria, part of an ongoing effort to improve relations between their nations.

According to the prime minister's science adviser, who related the scene to me years later, Gandhi told Reagan he needed a supercomputer for monsoon forecasting.

"Monsoon," the president said, "what's that? Big rain?"

"Yes, big rain," Reagan's national security adviser told him.

"And what is a supercomputer? That's a big computer?"

"Yes, a big computer," said the adviser.

At this, according to the version of the story I heard, Reagan thought for a moment, seemingly pondering what harm there could be in selling India a big computer to forecast big rain.

"Okay," he finally said.

Of course, Reagan's national security council and other federal agencies didn't think it was okay at all. I'm told that when the president returned to Washington, everyone tried to dissuade him from approving the sale. But it was too late; Reagan had promised, and he was intent on keeping that promise.

Eventually, the United States and India negotiated a deal. India could buy a supercomputer, a Cray X-MP/14, if it agreed to a few humiliating stipulations. First, it would need to pay full list price—about ten million dollars. Second, the facility that housed the computer had to be heavily secured, with soaring fences and guards holding long rifles. Finally, the only people allowed in the room with the computer were Cray employees or American citizens. No one of Indian origin would be allowed near the powerful machine.

No one except me, it turned out. As soon as the deal was struck, India's new secretary of science and technology, Dr. Gowariker, requested that I be put in charge of implementing the end-to-end global data assimilation and prediction model for weather forecasts. Cray was on board with my appointment, wanting an Indian scientist to serve as an intermediary between its employees and staff at the newly established National Centre for Medium-Range Weather Forecasting in Delhi. The United States was on board with my appointment because I had held a high-ranking civil service job and was now working at an American university. There was just one problem—no one wanted to pay my salary and travel costs.

The University of Maryland, happy to provide a very public favor to the government of India, agreed to keep paying my salary during the year it would take to install and operationalize the forecast system as long as I found substitute teachers for my courses and continued supervising my Ph.D. students. But there was still the matter of travel costs, the monthly flights and many nights in hotels the project would require. Knowing that a supercomputer could completely transform weather prediction—and therefore people's lives—in my home country, I was determined to make it happen, no matter the financial cost or logistical nightmare it posed. Together, Anne and I decided we would rent out our Maryland house and that she would spend the year in Denver with her family and our two small children. I went to the World Climate Research Programme (WCRP) and NOAA and managed to secure funding for travel.

In early 1988, I took the first of the fourteen flights to India I would take for the project, arriving in the capital city on a cloudless morning. I was so excited that I felt no jet lag at the end of the long flight. (That was not true for the future flights. Every month, I spent two weeks in Delhi, one week at the University of Maryland, and one week with my family in Denver—I was perpetually jet-lagged!) I had to get past the security guards at the gates of the new center in Delhi and then a small platoon of armed security guards, but I finally made it to the inner sanctum of the building. In that cold, windowless room, India's very first supercomputer buzzed, as big and ornate as a circular couch in a hotel lobby. Two security guards with long rifles stood at the door to the room. For a moment I stood in silent wonder before the machine, marveling at its journey and mine. A massive computer shipped from Minnesota and a boy from the village were about to change India forever.

For more than a year, I traveled back and forth between time zones, hemispheres, and cultural identities so often, I sometimes forgot where and who I was. On flights to Delhi, I traveled with software in my carry-on bag, the magnetic tapes that would supply the computer with the codes for data assimilation, a global atmospheric model, and a program to analyze its forecasts. I also helped with the recruitment of a team of young and talented Indian scientists and brought several colleagues from COLA, professionals who knew much more than I did about modeling and coding. Together we taught our Indian counterparts to use the computer, run the model, and make routine forecasts. After eighteen months, our work was finally complete. We had implemented a fully operational, modern numerical weather-prediction system for India.

When it was time to run the first experiment on the model, I knew exactly what I wanted to do. To satisfy my own curiosity, I had the scientists in charge run the model once using observations from Indian weather stations and then a second time leaving everything else the same but removing the Indian data altogether. The result? The forecast was more accurate *without* those observations. Now that the supercomputer was installed, the other scientists from India and I began our campaign to improve the quality of upper-air observations.

In March 1989, a large ceremony to celebrate the country's first supercomputer was held in Delhi. Prime Minister Rajiv Gandhi attended and made a short speech. I still remember his opening words:

"I would like to thank one particular person today, without whom this project would not have been successful," he said, looking out at the crowd from behind the lectern.

At that moment, all the spines in the room straightened, as everyone assumed the prime minister was about to utter his or her name. In addition to many scientists, there were a large number of people there who were involved in the purchase of the supercomputer, in transporting it and creating the air-conditioning and power infrastructure it needed. And there were a lot of wilted faces when Gandhi finished his thought: "That person is Ronald Reagan."

As a thank-you for my hard work, I received a handshake from the prime minister. But knowing that India finally had one of the most advanced weather analysis and forecasting systems in place and the capacity to issue more accurate weather forecasts to the Indian people—and knowing that I had had something to do with that—was all the reward I needed.

As for bringing seasonal prediction to India and the rest of the world, back in Maryland, the Center for Ocean, Land, Atmosphere Interactions was well on its way.

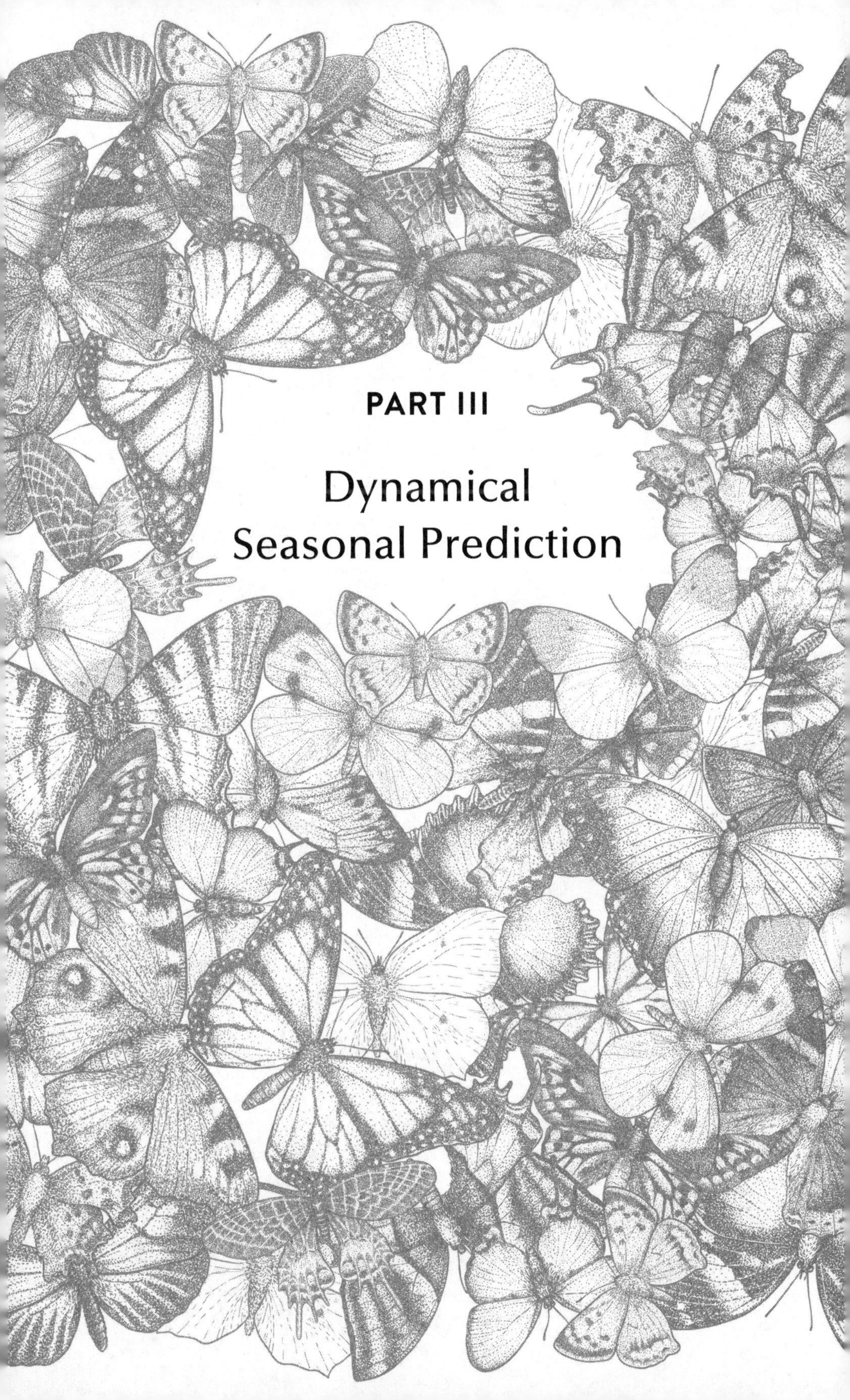

PART III

Dynamical Seasonal Prediction

Eleven

For ten years, from 1983 to 1993, the scientists of COLA made significant, often groundbreaking contributions toward establishing the scientific basis and feasibility of dynamical seasonal prediction using global models, investigating the potential to peer months into the future to foresee long-term, wide-ranging weather patterns. But now, instead of plugging exaggerated numbers into our models—turning the world into a parking lot or dumping ice cubes into the Arabian Sea—we were performing realistic simulations with realistic values, such as those of past observations, pushing the science into the real world where it could effect real change.

It was—not to be cliché about it—a dream come true. I was working alongside the smartest people I knew, advancing science I believed in, and doing it in the name of making the world a better place. It was the very thing I had had in mind during the walk through my drought-stricken village all those years ago as I considered whether or not to take the job in Pune.

⁜

COLA had been around for a just a year or two when my old friend Yale Mintz told me about a young scientist from England he had just recruited to NASA. Piers Sellers earned his doctorate from the University of Leeds in biometeorology—the interaction between land, and all its various forms of life, and the atmosphere above. Mintz the matchmaker had us both over for lunch soon after Sellers and Mandy arrived from the United Kingdom. It didn't take long for me to lure him to COLA; he came right after he completed his postdoc fellowship at Goddard Space Flight Center. That our town houses were within walking distance made us even faster friends.

I immediately saw that Sellers's work would be an important asset to our mission of using realistic models of the oceans, land, and atmosphere in dynamical seasonal prediction. The young scientist was interested in producing a physically realistic biosphere model, which he called a simple biosphere (SiB) model. The model would account for all the biophysical phenomena and energy fluxes happening on and under the ground, such as respiration, evaporation, and photosynthesis. By turning the whole world into a parking lot, Mintz and I had demonstrated that land surfaces played a dramatic role in seasonal climate, but our models were woefully behind in accurately representing what the land surface actually did to the atmosphere. A model like the kind Sellers wanted to build would be essential to producing accurate predictions.

At that time, even the best weather models in the world treated the land as if it were, in the words of my old mentor Suki Manabe, a bucket full of dirt. Sometimes it rained and the bucket filled with mud. Sometimes it didn't rain and the dirt in the bucket went dry. Evaporation from the bucket depended upon how wet the dirt was. It was, at best, an unsophisticated way to model the land, neglecting to account for all of the processes well known to biologists that seemed almost impossible to extrapolate to a global scale. Scientists could perfectly describe, for example, the photosynthesis of a single leaf, the way its microscopic pores sucked up carbon dioxide and exhaled water vapor. (This might seem like a minuscule thing, but when it's growing, a leaf can transpire many times more water than its own weight. An acre of corn? More than three thousand gallons of water each day.) But how could our models account for the trillions and trillions of leaves photosynthesizing every day, all the time, all over the world?

If anyone could help solve this massive problem, it was Piers Sellers, who seemed preternaturally drawn to risk-taking and adventure. Once, when Sellers was traveling to Colorado, I arranged for him to meet a friend of mine in Boulder who could take him skiing for the very first time. After the visit, my friend called me up. "It was really interesting," he reported. "Piers always chose the most dangerous trail."

So when Piers eventually confided to me that his ultimate career goal was to be an astronaut, I wasn't too surprised. Already an accomplished pilot, Piers asked if I could use some of the COLA's grant money to get him into the sky to observe the land-surface conditions in Maryland. A few days later, my neighbors experienced a terrible fright as Piers whizzed past in his small rented plane, barely clearing the tall oak trees next to my house (I had asked him to take an aerial photo of the area for fun).

Flying, in fact, played an important part in Sellers's simple biosphere model project, which inched closer to reality after a colossal field experiment he finally pulled off in 1987 in Kansas. It was designed to carefully measure what actually happened at the surface of one highly vegetated 225-square-kilometer parcel of tallgrass prairie. I still treasured my memories of flying over the stormy Bay of Bengal, and I was just as thrilled to sit aboard a research aircraft with Piers, a solid black globe with a measuring device attached sitting in front of him. This device calculated blackbody radiation, a way to create a scale of other radiation measurements taken from the same aircraft.

Sitting shoulder to shoulder with Piers, I was struck by an overwhelming sense of gratitude. More than ten years my junior, Piers represented the next generation of scientists. How lucky I was to be privy to his pioneering brilliance, to watch in real time as the ideas about the importance of land-surface processes I had formulated at MIT and NASA were elevated, refined, and made even more useful for society. To see this work carried out and supported by our center added to the thrill.

The field experiment made use of no fewer than five aircraft. The largest and most high-flying was the NASA C-130, which soared thirty thousand feet over the green rolling hills of the Konza Prairie measuring surface radiance. Below the C-130 were three flux aircraft that flew at decreasing altitudes measuring turbulent fluxes of CO_2 and latent heat. Last and lowest was a Huey, measuring radiance and soil moisture from close range. This scientific army of aircraft flew over grass, thickets of trees, creeks, and ponds at all hours of the day.

Meanwhile, on the ground, sixteen automatic meteorological stations and eighteen flux-measurement stations recorded data about meteorological and biophysical phenomena taking place at the surface. Scientists wandered around meticulously carving the land into strata of relevant information. They dug six layers deep into the wet earth to study root systems, poked around in the nodules of ancient tree trunks to figure out how much water the trees could absorb, inspected the canopy of a forest to learn how much rain it intercepted and how much water evaporated from the trees, grasses, and the soil below.

It took four days and four hundred hours of flight time, but we were finally able to collect enough data to tune our models so that we could do away with the notion of the ground as a bucket of dirt.

In reality, the land was a great, heaving thing, not separate from the atmosphere but in an endless dialogue with it, exchanging water and heat and gas and momentum. We also found, encouragingly, that data taken during our field experiment correlated strongly with the satellite observations we were receiving (and that would be increasing as NASA's Earth Observing System progressed through the end of the decade). In short, we had proven we could trust the satellite data even for areas over which such ground truth data was not available to validate the satellite data.

Using the tons of information collected from the field experiment,

Sellers, Mintz, Yongkang Xue, Nobuo Sato and their collaborators further refined their simple biosphere model, which translated all of this new knowledge into equations that could be integrated into our current general-circulation models.[1] (In the years to come, scientists continued to update the SiB to include improved treatment of seasons, snow, crops, carbon cycling, and so on.) At COLA, we wasted no time in coupling the SiB and our model, equipping us to better study the *L* in our name. It wouldn't take long for it to play an integral role in the creation of several international research programs, and it helped institutions all over the world improve the treatment of land in their models to address land-use problems, predict droughts, and avoid environmental degradation.

As for Piers, his dream to go to space finally came true in the 1990s, when he joined NASA's astronaut corps. Sellers flew three missions and helped enhance the International Space Station, spending thirty-five days in space and more than forty hours outside the shuttle on space walks. The most dangerous trail indeed.

Sellers didn't forget his climate comrades back in Maryland, though, taking with him on one of his missions a bright blue flag with the COLA logo signed by each of his former colleagues. Today that flag hangs in the COLA conference room near my office, where I see it often and think of Piers—and the swaying tallgrass prairie.

While Piers was working to make his weather-model and space-flight dreams come true, COLA spent those same years lobbying for another one of my passion projects.

I'll never forget the morning in 1984 when my taxi pulled up to the gate at the European Centre for Medium-Range Weather Forecasts in Reading just as Anthony Hollingsworth, a rosy-cheeked Irish scientist and head of the center's data-assimilation group, also arrived. As I stepped out of the car, Tony glanced over, but he didn't give me a warm smile or a friendly wave. Instead, I watched as his face fell into something of a frown.

"Uh-oh," he said, deadpan. "Here comes trouble."

By the 1980s, I had earned a reputation for being a bit of a pest. Back then, at the same time I was bugging every Indian official I met to invest in modernizing the country's forecasting system, I was also annoying my Western colleagues with another (I thought) game-changing idea that I was having no luck in getting off the ground.

It had come to me during my NASA days in the early 1980s, when the agency launched its Global Habitability program (eventually renamed the Mission to Planet Earth) to investigate how our planet's climate was changing.[2] From probes sent to Mars and Venus, NASA scientists had found plenty of evidence that planets like ours could and did undergo extreme climatic changes; after all, rivers of liquid water used to flow across Mars, and for billions of years, Venus had surface temperatures comparable to the ones on Earth. What's more, in the middle of the last century, geologists began to see that Earth's climate had shifted more rapidly than anyone had previously believed. Evidence in ice and sediment cores proved that our planet had experienced transformations that were both severe and surprisingly rapid.

The idea that Earth's habitability could change in one person's lifetime made understanding our climate and how it might be changing an urgent task, and NASA asked its scientists to come up with new ideas on how we could do that.

As a meteorologist working on dynamical seasonal prediction and, in particular, monsoon forecasting, I thought first of the dearth of historical climate data, especially in tropical regions and the Southern Hemisphere, areas where there were few weather stations. Numerical weather-prediction centers of the world routinely analyzed the global observations using the best model available at that time. Almost on a daily basis I saw my supervisor, Milt Halem, and my good friends Eugenia Kalnay, Bob Atlas, and Joel Susskind discussing how to assimilate satellite-derived temperature in data assimilation, the biggest challenge of the era. For decades now, our models and data-assimilation methods had been improving. It occurred to me that if we fed all of that old and incomplete data—including data that had arrived late and had never been used—into the most up-to-date weather-prediction models programmed with the latest data-assimilation methods, we could generate a fuller and more accurate picture of the four-dimensional structure of the atmosphere.

My pitch was to start with ten years of data. Using the most recent model and data-assimilation method, we would take all of the initial observations ever gathered for weather forecasting and use them again to produce new and more accurate initial conditions for every day, every hour, and every place on Earth. A whole decade of improved and internally consistent data (consistent because the same model and the same assimilation method would be employed for the whole period) was integral to figuring out how the atmosphere worked and how the climate might be changing.

Everyone I told about reanalysis—as I took to calling it—thought it was a great idea. That included my colleagues at NASA and people at the National Science Foundation. Jay Fein of the NSF was the first to fully endorse the idea of reanalysis, and he encouraged me to pursue it further. I also won support from a major El Niño research project (more on that later). There were just two problems, and they were big ones. First, nobody was sure all that past data could even be located. Second, the effort would require significant money and staff time, and no one was willing to do it. Admittedly, toiling over years-old data wasn't the most exhilarating idea I'd ever come up with—but I felt sure it was one of the most important.

The first US agency I approached was the National Meteorological Center (now called National Environmental Prediction Center), part of NOAA. After some hemming and hawing, the director eventually turned me down, saying, "We're in the forecasting business; we don't want to be looking backward." Even when a National Center for Atmospheric Research scientist, Roy Jenny, told me that he had the data, that he had been meticulously saving it on tapes and disks for years, and that he would be happy to give it to me, the answer from the leadership was still no.

Undeterred, in 1984, I flew to London and took that taxi to the

ECMWF. I met with Dr. Lennart Bengtsson, the director, and his two most senior scientists, Dave Burridge and Tony Hollingsworth, the man who had been none too happy to see me (a future friend, I should say). But even he couldn't deny the idea was sound, though all three men expressed their concern that it would take a huge amount of staff time to gather the past data and perform all the tedious but necessary quality-control checks. Dave quipped that we'd have to put the staffers in a windowless office. I didn't understand the joke. "Otherwise they may lose the will to live and jump," he explained. Everyone shared a laugh except me.

Finally, Lennart spoke. The European center would commit to trying reanalysis.

In science, *yes* is merely the *beginning* of the battle. Even with Lennart on our side, we first needed to write a proper scientific paper describing the merits of the idea and the potential problems we might run into.[3] I invited Lennart to come to Maryland and write it with me. In the paper, which was published in the *Bulletin of the American Meteorological Society*, we argued that "such data sets will be quite useful for studying global climate change." Although the idea became widely accepted, it again lay dormant, stymied not by a lack of enthusiasm but by a lack of funding. But then I found myself the director of COLA—and suddenly I had access to both staffers and funds. I asked my staff if they were on board with allocating money and time to run a proof-of-concept pilot reanalysis experiment using just fourteen months of data. They were.

Just before we got started, I attended a meeting of the National Research Council in Washington, DC. A NASA scientist whom I had known from my Goddard days and who was now a senior program director in charge of supporting research projects proposed by university scientists was also there. He asked me if we could meet during the lunch break, but before the meeting adjourned for lunch, he passed me a note informing me that he had about two hundred fifty thousand dollars to spend on new and innovative projects. Needless to say, this has happened to me only once in my life.

When we met during the break, I had a concrete idea to propose: reanalysis. He liked it enough that he asked me to send him a formal proposal. As soon as I returned to the office, Jim Kinter and I started working on preparing one. We submitted it, and not long after, we learned that on top of our existing grant funds, COLA now had an additional quarter of a million dollars for its pilot reanalysis. In early 1989 we organized a meeting of modeling groups from NOAA, NASA, NSF, and ECMWF to discuss plans for a ten-year reanalysis. All the groups at least agreed to save their quality-controlled observations so that reanalysis could be done at a later date.

The one-year pilot by our group finally confirmed that, yes, reanalysis could be done. Soon every major weather-prediction center in the world was performing its own reanalysis efforts, one of them, the ECMWF, supported by one hundred thousand dollars from COLA itself. That's how strongly we believed in the importance of reanalysis.

After ECMWF's twenty-year reanalysis came NOAA's, led by a former Charney student and the first woman to earn a Ph.D. in meteorology from MIT, my friend Eugenia Kalnay. While her predecessor had no interest in "looking backward," as he'd put it, when Eugenia became the director of NOAA's Environmental Modeling Center, she made reanalysis an immediate mission. (Kalnay was always a bit of a rebel; in 2011, we would march together at an Occupy Wall Street demonstration in DC.) Under her leadership, the staff at the center gathered all the global observations of every day for the past forty years (which were archived at different institutions around the world), plugged them into the most up-to-date and advanced data-assimilation systems and weather-prediction models available, and reran the models in six-hour intervals, just like meteorologists do every day at numerical weather centers around the world. In other words, Kalnay and her staff prepared a daily weather report for every day of the past four decades.

In 1996, Kalnay published a paper (titled, simply, "The NCEP/NCAR 40-Year Reanalysis Project") that included the reanalyzed data, a gold mine for scientists working in the field. It quickly became—and remains—one of the most cited papers in the field of meteorology. Soon other countries embarked on their own reanalysis projects.

These reanalysis products have become an indispensable source of data for diagnostic studies of the present and past climate. I once sat in a meeting and heard a fellow scientist say that reanalysis was among the greatest accomplishments of our field (along with weather and climate models). I didn't say anything when I heard that, just smiled to myself. Sometimes the best ideas aren't the boldest, the most exhilarating, or the easiest to sell. Sometimes it pays to be a pest.

The simple biosphere model and reanalysis were enormous accomplishments, but they were perhaps overshadowed by one of the longest and most important projects we took part in that decade: the Tropical Ocean–Global Atmosphere (TOGA) program, which set out to observe, understand, and predict El Niño. Although my research had demonstrated that there is a scientific basis for dynamical seasonal prediction (which was by now widely accepted) and several countries had started making dynamical seasonal predictions using observed ocean initial conditions without changing them—called persistence—for a season, no one had a global coupled ocean-atmosphere model that could predict *future* ocean conditions. After TOGA, climate modeling research culminated in the first seasonal forecast ever made with a global coupled ocean-atmosphere model—the realization of my nearly two-decade-long dream.

The 1982/1983 El Niño—the one that caught the world, including its top climate scientists, off guard—was among the most destructive in history, responsible for as many as two thousand deaths and thirteen billion dollars in damage.[4] It was not predicted because El Niño forecasting did not exist back then; in fact, it wasn't even *detected* until its peak, since real-time observation data in the tropical Pacific was difficult to come by, and the satellite data turned out to be erroneous thanks to a slew of technical problems and cloudiness.

From my perspective of dynamical seasonal prediction, that El

Niño provided a unique opportunity. I wanted to know if the existing global atmospheric models could replicate the most important aspects of what had happened around the planet if they were given the actual observed tropical ocean temperatures—gleaned from routine shipping and satellite data—from the 1982/1983 El Niño. Until then, all the modeling experiments had been done using idealized and artificially created ocean temperatures, but now we finally had good data for an actual El Niño. If the models could replicate the major global effects of El Niño, and if we were able to predict El Niño itself, it would enable society to prepare and save untold lives and livelihoods. So I asked several modeling groups around the world to run their models with the actual tropical ocean surface temperatures during 1982 and 1983.

In May 1984, with support from the World Climate Research Programme, I convened an international workshop of nine modeling groups in Liège, Belgium, to present and discuss how accurate their models had been in reproducing the observed tropical rainfall and global atmospheric-flow patterns. This was the most exciting and stimulating meeting I had ever organized. We were about to find out what these global atmospheric models were made of. It also became a bit of a competition between groups and their models. (An article about the meeting in *Science* described the workshop presentations as "a real shootout, a grand show and tell.") It was gratifying to learn that most (but not all) of the models successfully captured the observed changes in tropical rainfall when they had been fed the actual tropical ocean temperatures.

This was the first demonstration, using global atmospheric models and actual ocean temperatures, of the idea that tropical ocean surface temperatures strongly influenced rainfall and circulation, a hypothesis I had put forward more than ten years ago. It was a fantastic feeling to see my long-ago optimism validated. (At the same conference, George Philander and Anne Siegel from Princeton showed that the Geophysical Fluid Dynamics Laboratory–coupled model could simulate El Niño–like phenomena, although it still couldn't predict El Niño.)

The workshop provided a major boost for TOGA, which the

World Climate Research Programme and the US National Research Council launched in 1985. In the tradition of Charney's Global Weather Experiment, TOGA would establish an unprecedented ocean-observation system in the tropical Pacific and unleash a new era of coupled ocean-atmosphere models.

The cornerstone of the observational component was the Tropical Atmosphere Ocean Array, which consisted of seventy moored buoys stationed along the equator that measured surface wind, sea-surface temperature, and deep-ocean temperature and transmitted that data to shore in real time via satellite. In addition, the new observation system relied on drifting buoys, tide gauge stations, and volunteer ships. This elaborate setup allowed scientists, for the first time, to know what was happening in the vast Pacific Ocean at practically any moment of the day. El Niño would never sneak up on us again.

But this was only one goal of TOGA; the others—to understand El Niño, model it, and predict it—were a bit more complicated. There was not a single global coupled ocean-atmosphere model anywhere that could do what Charney and von Neumann had done for weather prediction: start with the observed initial conditions of the atmosphere and oceans and predict the evolution of that coupled system for the next one or two seasons. TOGA funded modeling groups around the world to develop coupled ocean-atmosphere models capable of simulating and predicting El Niño.

COLA was one of those groups. For ten years, our scientists, led by Ben Kirtman, labored over the model, performing the same fine-tuning that an NWP model requires but with atmosphere *and* oceans. By 1995, the research community, including our team at COLA, had the scientific and technical expertise to begin making dynamical seasonal predictions regarding the tropical Pacific Ocean and its atmosphere. (It was not until 2006 that NOAA started making dynamical seasonal predictions using global coupled ocean-atmosphere models.)

To celebrate the end of one of the most successful climate programs in history, a grand toga party was held in Melbourne in a cavernous building that used to serve as the city's jail. Several TOGA scientists (including the author) came to the party wearing bedsheets.

Two years later, in 1997, I had the privilege of addressing the global climate community at the WCRP conference in Geneva. I was asked to present a report on the status and accomplishments of TOGA; I could not have been given an easier task. That spring, climate models had predicted that a major El Niño would occur by the end of the year. Thanks to the new observation array and our new coupled models, I had the necessary data to declare at that conference that ocean temperatures in the eastern Pacific in July and August of 1997 were the warmest ever observed in the past one hundred fifty years. Better yet, I had enough confidence in our model to predict that the ocean temperatures would be even warmer six months later. *This* was what I had been imagining all those years ago back at MIT when I learned that, thanks to the butterfly effect, seasonal prediction was an impossibility. It had taken more than twenty-five years, but up there at that podium, I was essentially living out a dream: giving society fair warning that powerful, life-threatening weather was on the way.

My audience understood the gravity of the moment—they responded with pin-drop silence. A new chapter in dynamical seasonal prediction had been written. It was also a momentous occasion in my own career. Soon after the successful prediction of the 1997/1998 El Niño, Ants Leetmaa, head of the climate prediction group at NOAA, started a lecture in Miami with this joke: "I have heard a rumor that Shukla would retire only after the dynamical seasonal prediction problem is solved. Perhaps it is time for Shukla to think about retiring?" After all, what else would a scientist obsessed with seasonal prediction have left to do?

As it turned out, a lot. While the 1997/1998 El Niño became a shining example of the success of TOGA, its prediction had some unintended consequences. It gave a false impression to some in the forecasting community, and especially to the research managers of NOAA, that the problem of El Niño prediction had been solved. This was very unfortunate because it led to drastic reductions in funding for research on the theory, modeling, and predictability of El Niño. The El Niño predictions have never again been as accurate, and since 1997, several El Niño forecasts have failed.

Moreover, even though models are able to predict ocean temperatures for strong El Niños, prediction of global rainfall and circulation remains a challenge. Even in 1997, although the prediction of tropical ocean temperature was quite accurate, predictions of the global circulation over some parts of the world especially over India were glaringly wrong.

By then, we were well aware from past observations that intense El Niños produced severe monsoon droughts over India, like the 1972 drought I had experienced while visiting my village during graduate school. When the Indian authorities asked me confidentially for my opinion about the 1997 monsoon season for India, I confidently supported the conventional wisdom—confirmed by the models—that the 1997 monsoon season would be a drought year for India. Well, nature has its own way of keeping scientists in their place—I was wrong; we all were, and 1997 was a normal monsoon rainfall season for India.

Some suggested Lorenz's butterflies were at work again, fiddling with our hard-won model. Some research indicated that the influence of El Niño was neutralized by the Indian Ocean temperatures, which were not correctly predicted in 1997. Over the years, several Ph.D. dissertations and a large number of scientific papers have been written about why the dog did not bark. The only thing we know for sure is that there are aspects of the climate system that remain a mystery even now. There is much more work to do.

Still, I'll never forget standing behind that podium in Geneva in December 1997 and alerting the world that El Niño was coming. One of my COLA colleagues had suggested that I should carry copies of my slides to that meeting so that participants from developing countries—the ones that would be most affected by El Niño—could take home copies of my presentation. I could fit about a hundred in my bag. At the end of my lecture, the chairman announced that they would be available on the table near the podium. What followed was a mad rush, and my copies were gone in less than a minute.

That evening and the next morning, I must have been interviewed by fifteen television news reporters, and Anne told me that my warning about the warmest ocean temperature in one hundred fifty years was the opening story on the six o'clock national news. Hours later I boarded a plane for Delhi, and when I reached my hotel, I flipped on the TV to see if my fifteen minutes of fame included international coverage. Alas, other events in Paris that night had overtaken the El Niño story, and my time was up.

MORE ON MODELS

The success of TOGA was due to thousands upon thousands of hours of work refining, tweaking, and tinkering with the climate models that are now the beating heart of climate science. But a lot of professions rely on models. Artists use models to make their paintings as lifelike as possible. Engineers build models, miniature representations of structures to be built, to test or sell their designs. For physicists and mathematicians, a model is a system of equations that describe the properties of the processes that the scientists wish to study. In my business, weather and climate science, a model is a system of mathematical equations for the laws of physics that govern the behavior of the real weather and climate systems. Models are arguably the most important tool for climate scientists. But how did that come to be?

In the 1940s, simple models representing a single layer of the atmosphere were used to predict large-scale wind patterns. In the late 1950s, Charney and von Neumann revolutionized weather prediction by using more complex equations and a fast digital computer. With faster computers and improved understanding came the ability to model the physical processes of convection, radiation, boundary-layer dynamics, and orographic influences. Weather prediction using complex global models has become a routine operation for the weather services of the world. Countries that do not have trained scientists or the resources to acquire high-speed computers either use regional versions of global models or simply take forecasts from other global models.

The computer outputs of weather-prediction models give quantitative estimates of weather variables at the resolution of the grids used by the models. These huge volumes of data coming out of the computers need to be summarized in a language that is easily accessible and usable by the public at large. Traditionally, this has been done by radio and TV weather reporters. However, with the advent of social media and handheld devices, there are now nearly ten thousand weather apps that routinely provide local weather information. The fundamental equations that produce the weather forecasts are the same; it is the details of the resolution, the treatment of physical processes in the model, and the manner in which the predicted weather conditions are interpolated or calculated at fixed latitudes and longitudes that make the forecasts different. There is a proliferation of weather apps because different weather apps use different methods to process, to interpolate, to distribute, and to display the weather forecasts.

Climate models are far more complex than weather models. Weather models are limited to the modeling of atmosphere only, but climate models include the atmosphere, oceans, and land-surface processes—and the interactions among the three.

Models for global atmosphere and oceans are referred to as the atmosphere general circulation models (AGCMs) and ocean general circulation models (OGCMs) respectively. When an AGCM and OGCM are dynamically coupled to each other, the new model is referred to as a coupled general circulation model (CGCM). Suki Manabe and Kirk Bryan of GFDL were the first to develop a coupled ocean-atmosphere model. These coupled models make regular seasonal predictions, but their most significant application over the past thirty years has been estimating the future projections of climate change due to increased emissions of greenhouse gases by human activities.

Currently, there are more than twenty-five modeling groups in the world running climate models, all at different resolutions and with different treatments of physical processes like convection and cloudiness. That is why different models give different projections of climate change. Different models have different strengths and weaknesses, and the modeling community has not been able to develop a consensus on quality-control measures to reject the models that do not meet a minimum standard of performance in simulating the current climate. Following the principles of "model democracy," the climate projections made by the modeling groups of *all* nations that contribute their results are announced every five years by the Intergovernmental Panel on Climate Change.

For the past fifty years, a major challenge for the climate-modeling community has been correctly simulating the past climate, so a major focus of climate-modeling research is improving the fidelity of the coupled models in simulating the observed climate. Once the modeling groups are satisfied with the fidelity of their models, they use them to understand the properties of the climate system, predict global patterns of circulation and rainfall, and make future projections of climate change.

At first, these global coupled models could not correctly predict the future evolution of ocean conditions from an observed ocean state. Stephen Zebiak, a Ph.D. student of Mark Cane at MIT, developed a simple coupled model of the tropical Pacific Ocean, sometimes affectionately referred to as a "toy model for El Niño," that was the first to make skillful predictions of the El Niño–related ocean-surface temperature changes over the tropical Pacific. The groundbreaking work by Cane and Zebiak pushed the climate-prediction community to improve global coupled models and ocean data-assimilation methods. It took nearly ten more years for the global coupled ocean-atmosphere models to predict El Niño as well as the toy model did. The global coupled models are now used routinely to predict global patterns of rainfall and atmospheric winds.

Current climate models consider only the physical processes in the atmosphere, on land surfaces, in oceans, and on ice. But since the chemical and biological components of the Earth's system are intimately linked with the physical climate system, not to mention human interactions, it is important that models for the future projections of climate include interactions with them both. These are called, naturally, Earth system models. Research scientists are already working on the development of interactive earth system and human system models.

For example, as the climate changes, it will affect soil hydrology, plant processes and vegetation dynamics, snowpack and permafrost, glaciers and ice sheets, and terrestrial and ocean biogeochemistry, all of which in turn will affect the carbon and nitrogen cycles in the atmosphere. Earth system models must take into account the interactions among physical, chemical, biological, and ecological systems. Developing and validating such a complex model will be a herculean challenge for the scientific community—after which we can begin to estimate the *predictability* of the Earth system. We have only just begun to try.

Twelve

Professionally, I was doing very well. The Center for Ocean, Land, Atmosphere Interactions was dedicated to advancing seasonal prediction—and succeeding—while I had the opportunity take part in projects all over the world and collaborate with some of the brightest scientists in the field. The 1980s was a frenzied, jet-lagged decade.

Anne was working harder than I was, raising our two children, Chandran and Sonia, largely on her own. Chandran, named after my father, was born in 1981, and Sonia followed in 1983. They brought tremendous joy into our lives. But while I was flying to England, Switzerland, Italy, and India, they started walking, started kindergarten, started to forget who I was.

Once, when I had been gone for a particularly long stretch, Sonia told Anne, "I forget how Dad looks." To help her remember, Anne took our daughter to my closet, pulled out one of my black blazers, and held it up to conjure my presence. I believed fully in what we were doing at COLA, but I also knew that my children paid a heavy price for that work.

Parenting in person did not come naturally to me. During my first marriage, I did not even learn my wife's name (until I needed it for paperwork in Pune), did not meet my daughter until she was a toddler. In an Indian village, men were basically sperm donors; when I was growing up, I almost never saw fathers—including my own—interacting with wives and children (especially daughters). A man's role when the children were born was to provide financial and material security the best he could. If that meant spending the majority of his time away from the family, so be it.

With Chandran and Sonia, though, I wanted things to be different. My relationship with Anne had shown me another kind of marriage, and I was eager to try a new way forward with my children. From the beginning, I was as involved as I could be, attending Lamaze and birthing classes when Anne was pregnant with Chandran. These classes made me a nervous wreck! Who knew that so many things could go wrong? Months later, when Anne yelled down the stairs that her water had broken, I was so on edge that I forgot to put down the glass of cranberry juice I was holding before I ran up to take care of her. For months, the ruby stains on the stairs were a souvenir of Chandran's arrival.

In between business trips, I loved to play with them, to kick around the soccer ball in the garage, referee their squabbles, and spoil them in the little ways I could. Anne and I took our two toddlers, who were sometimes fussy little kids, all over the world; we were once kicked out of a restaurant in Paris, France.

We also took them to India as much as possible. The first time Chandran endured the fourteen-hour flight to India, he was only seven months old. We stayed for half a year, and our son learned to crawl on the dusty floors of a guesthouse in New Delhi. When Chandran was eight and Sonia six, we took them to Varanasi, the city on the banks of the Ganges where my father had died and where I had gone to college. Chandran was so fascinated by the snake charmers in Varanasi that he asked us to buy him a turban just like the charmers wore. Five years later, we took the children to visit my village for the first time. It seemed all of Mirdha turned out to welcome them, impressing the kids with a small band of horses and even an elephant, and a palanquin for Anne.

When Chandran was fourteen, I took him to the Olympics in Atlanta, my son lugging a number of long-lensed cameras to practice his action shots. Chandran was always fascinated by gadgets and machines, while Sonia accrued friends with ease. It was a pleasure to watch them grow into such unique young people. Often, I thought of all the fathers I knew in India who, for cultural or personal reasons, could not know the pure joy of witnessing their children's childhoods.

It made me that much sadder that I had missed Pooja's. I mostly knew her through yearly visits and letters. I had made sure that she attended the best boarding schools and I sent as much money as I could back to India to care for her and the whole family, but knowing that my daughter was halfway around the world was a shadow that hung over all the good things that happened during that time.

I was happy and I knew I was so lucky, but sometimes my life felt like an endless tug-of-war between cultures, between countries, between all the people I wanted and needed to care for. My days were filled with so much richness—my wife, my children, meaningful work—and also echoes of loss, of family that was too far away and dear friends I'd never see again.

Just before my son was born, Jule Charney called me from Boston. This wasn't unusual, but the urgency I heard in his voice was. He told me he would like me to come visit, and I said I would look at flights. The truth was I was so busy with work and preparing for a new baby, I wasn't sure when I would be able to get to Massachusetts. But then Charney told me he had already looked at flights and had a certain one in mind, and I realized time was of the essence.

I'd known that my old adviser was sick, but I hadn't realized just how far his cancer had progressed. When he opened the door of his Cambridge apartment, I could hardly believe this was the same man I had met in Tokyo all those years ago. My vibrant, charismatic friend was shrunken and sallow, every muscle deflated, his eyes dark. Still, he gave me a weak smile and a weaker handshake and insisted on serving me lunch with shaky hands.

Over our meal, the last one we would share, Charney asked me for a favor. He was advising one final Ph.D. student at MIT, a Brazilian Earth system scientist named Carlos Nobre. In the event that Charney could not oversee Nobre's dissertation, he wanted me to take over as adviser. I told him I would. Charney had become a close family friend. On several occasions when he visited NASA, he'd stayed with us. During one of his visits, we played soccer in my front yard with his former students Eugenia Kalnay, Inez Fung, Michael McIntyre, and other scientists. In truth there was nothing that Charney could have asked of me that I wouldn't have done gladly. He was the one who had made me the scientist I was, who had taught me to be self-critical but bold, to say the big, crazy ideas out loud—a computer-derived forecast, a global weather experiment, long-term monsoon prediction, dynamical seasonal prediction. He was the one who asked the organizers of a major conference on the Global Weather Experiment in 1979 to invite me, and when the organizers said that the seats were full, he offered to give up his seat for me, and quickly the organizers found a place for me.

My father had taught me to swim by throwing me into a muddy pond, and like him, Jule Charney had taught me to take on seemingly unsolvable problems by simply jumping into the work headfirst. Everything could be linearized, could be made solvable, he said. Everything.

Several weeks after my lunch with Charney, his partner, Pat, called to tell me that he was once again asking for me, saying my name over and over in his delirium. A few days later, she called to say that I needed to get there if I wanted to say goodbye. Yale Mintz, Arnt Eliassen—a friend and longtime collaborator of Charney's from Norway—and I rushed to the airport and caught the next flight to Boston, but it was too late; by the time we arrived, he was gone, dead at only sixty-four. I have thought of him almost every day since, wondering what he would say about this problem or that experiment, wishing he could see how far we've come.

As it turned out, my last favor to Charney became a lasting friendship as well as a productive professional relationship. Carlos Nobre and I had a lot in common. While my childhood in India had left me with a deep fixation on monsoon prediction, Nobre's had left him obsessed with the Amazon rainforest and its impact on Brazil's climate, worried about what would happen if deforestation continued at the breakneck pace of the 1980s. After he finished his Ph.D., I invited Nobre to join COLA to study the impact of Amazon deforestation using our models.

By the late 1980s, there was burgeoning awareness and concern about the ecological disaster unfolding in the Amazon, where millions of trees were being felled to create wide swaths of land for crops and cattle pastures. Highways were built, mines dug, and dams constructed atop of the jungle, and all this activity was supported, if not subsidized, by the World Bank and the nation's government. Those who stood to make money from the relentless logging and the industries it created argued that the forest was a regenerative resource, that it could be used and regrown, but scientists like Nobre weren't so sure. By now we understood that the land surface didn't simply receive the weather; it also created the weather.

Nobre and I weren't the first scientists to wonder what would happen if the Amazon was chopped to pieces. We were, however, the first to have access to a remarkably sophisticated model that had been coupled with a realistic treatment of land, thanks to Piers Sellers. Because of his simple biosphere model, we were able to run an experiment in which we wiped the Amazon off the face of the Earth.

What we found was shocking.[1] When the Amazon was replaced by degraded grassland, the mean surface temperature over the Amazon surged by 2.5 degrees Celsius. Evapotranspiration fell by 30 percent and precipitation by 25 percent. In this scenario, trees shed their leaves to save energy, allowing more sunlight to reach the forest floor; this dried out the understory and created ideal conditions for wildfire. Less water in the soil meant less water for trees to soak up and for leaves to release back into the atmosphere. Less evaporation meant fewer clouds, fewer storms, and even less water for the soil—a vicious dry cycle called savannization.

Our simulation also showed an increase in the length of the dry season, a troubling result, as tropical rainforests can occur only when there is a brief or nonexistent dry season.

In short, runaway deforestation would eventually create a feedback loop of heat and drought from which the jungle would never recover. Countless plants, animals, and insects would go extinct. Instead of a carbon sink, the rainforest would become a carbon *source*. Our work was published in *Science,* and for weeks journalists were calling COLA scientists to interview them about the study, the first clear and reliable projection of the ominous outcome should the destruction of the rainforest continue.

It had never been a stretch to say that clear-cutting the Amazon was a bad idea, but what Nobre, Sellers, and I had done was demonstrate just how catastrophic it would be. The work also demonstrated what a powerful tool our model was and how long-term climate predictions could shape public understanding and policy regarding the natural world. Piers Sellers and Carlos Nobre laid a strong foundation for the *L* in COLA. (One of our Ph.D. students, Paul Dirmeyer, has made numerous outstanding scientific contributions in this regard and is now the leader of land-atmosphere interaction research at the center.)

A few years later, Yongkang Xue, one of the center's young scientists, and I worked on a similar experiment, this time focusing on *re*forestation using the COLA model. Xue, a recent graduate of the University of Utah, was interested in the crisis of desertification in the Sahel region of Africa, the same problem that had caught Charney's attention back in the late 1970s. Since then, the Sahara Desert had continued to creep south into the Sahel grasslands, claiming vital farmland and jeopardizing the food security of more than five hundred million people.

Desertification is one of the worst things that can happen to a landscape—and a society. Both the Sumerian and Babylonian Empires are thought to have collapsed due to desertification; their soil became unfarmable, game and water scarce. Even now, it is almost impossible for people to adapt when the land they live on dies.

Xue and I wanted to see if there was a way to reverse this process, or at least stop it, by planting vegetation at the southern boundary of the Sahara. If a loss of trees exacerbated droughts, then an infusion of them would do the opposite, we hypothesized. And so, once again relying on Sellers's biosphere model, we ran a simulation, this time turning the Sahel region green with trees. We found that aggressive reforestation could work, could put a halt to the devastating desertification spreading across the landscape.

The idea of reforestation took root—to use a pun—and today, efforts are underway to construct the Great Green Wall, a four-thousand-mile-long, ten-mile-wide barricade of vegetation that we hope will one day stretch across the entire continent, from Senegal to Djibouti, providing a formidable barrier against the encroaching Sahara and protecting the lives and livelihoods of one of the fastest-growing populations in the world.

Climate prediction is powerful; climate prediction in developing countries saves lives.

Providing support and opportunities to scientists in developing countries has been the mission of the International Centre for Theoretical Physics (ICTP) since it was founded in 1964 by Pakistani physicist Abdus Salam. Salam was something of a hero to me, a Nobel laureate who had made huge advancements in particle physics and who was committed to using that science to improve conditions in Pakistan and other developing countries. So dedicated to this mission, Salam never took a day off—not even for holidays—and rarely attended social functions.

I met Salam in 1988 at an ICTP meeting, where I was quite starstruck to find myself sitting next to him at lunch. A broad-shouldered man with gray hair, a gray beard, and square tortoiseshell glasses, Salam slid into the chair next to me and straightened his pinstripe suit. He glanced at my name tag, picked up his silverware, and said, "Tell me what you do, Dr. Shukla."

I told Salam about Maryland, about COLA, and our ever-growing scientific certainty that dynamical seasonal prediction wasn't just possible, it was imperative. "Our work has shown that in tropical regions, where most of the developing countries are, climate is more predictable than in extratropical regions," I said.

This got Salam's attention; he, more than anyone in the world, knew that advances in science and technology had largely benefited developed countries. He put his fork down gently, offered me a silence I interpreted as an invitation to speak more, and stared intently at the flowers in front of him while I spoke, as if what I was explaining—the influences of boundary conditions and the rotation of the planet—was written on the lilies' orange petals. "It's a pity, but those countries don't have the scientific capacity or the resources to take advantage of this unique gift provided by nature," I concluded. "And as you know, these are the countries that need it most, the nations where people rely almost solely on agriculture to make a living."

Salam beckoned to his deputy director, who was sitting at the other end of the long table. In a moment, his second in command stood before us. "This is Dr. Shukla," Salam said, gesturing toward me. "And we should start a climate group here."

A flash of panic appeared in the man's eyes, the look of someone who realizes that his life has just become much more complicated than it was only moments ago. But the man said only "Yes, sir," which was what everyone always said to Salam.

As dessert was served, I had to keep careful control over the corners of my mouth lest they rise up into an enormous, incredulous smile. I couldn't believe it; my hero was impressed by what I had to say! His organization, one of the most respected in the scientific community of theoretical physics and advanced mathematics, was going to start a group on climate to work on the very same problems I had been tackling for years. Lost in my own self-congratulation, I felt Salam's elbow gently nudge mine.

"You should come be the head of this group," he said in his thick Punjabi accent. "You better start planning."

I immediately thought of Anne and the kids, my colleagues at COLA, and all the plane tickets stacked up on the bureau at home; this was the year of fourteen flights back and forth to Delhi. In the weeks to come, Salam sent me an offer letter for the position, complete with a big salary, a car, and a chauffeur. I was flattered, but the answer was an easy no. Nothing could lure me away from my colleagues at COLA and the critical work we were doing. Along with my old friends Antonio Moura and Venkataramanaiah Krishnamurthy, I did help ICTP establish their climate group. Today this group, called Earth System Physics, conducts research and builds scientific capacity for weather and climate prediction in the developing world.

No matter how busy I was that year, one request I could not turn down was an invitation to speak in front of the pope at the Pontifical Academy of Sciences, an institution that has existed in one form or another since 1603 and whose first president was Galileo Galilei. The academy invites scientists from all over the planet and of all different faiths to discuss problems in math, science, and the physical world. One of the academy's areas of focus is science in developing countries; COLA's work on the Asian monsoon and the Sahel droughts was what got me invited to the Vatican. The symposium I was invited to concerned climate-driven crises, especially the historic and devastating droughts in Africa, one of the biggest disasters of the decade.

Now, no disrespect to the Vatican, but I've never been all that enamored of that institution. As a student of population dynamics, I very much believe in widespread access to and use of birth control. But the guest list of this symposium read like I had written it myself; it was full of friends and old colleagues, and there was no way I was going to miss this impromptu reunion.

The Pontifical Academy of Sciences is housed in the Casina Pio IV, a sand-colored patrician villa adorned with sculptural stucco-work that looms over an ornately tiled courtyard. The interior of the five-hundred-year-old structure is somehow more beautiful than the outside, practically wallpapered with frescoes by famous Renaissance painters. For three days, my fellow scientists and I wandered through halls dripping with iconography, their alcoves stuffed with statuettes. We dined beneath vaulted ceilings, served by waiters wearing tail-coats. Boxed lunches in a featureless conference room this was not.

On the final day of the symposium, we had an audience with Pope John Paul II, who sat before us cloaked in white. After all the scientists in the room had given presentations on their efforts to predict or ease drought on the African continent, the pontiff lectured us on the potential of climate control, which we found somewhat amusing. We listened politely; after all, His Holiness had direct access to a higher authority. If anyone could pull off weather and climate control, it was him.

At the end of the afternoon, the pope gave each of us a handshake and a priest next to him gave us a rosary as someone took a photo. When we returned to the hotel, we wondered if we would ever see those photos. Sure enough, soon we found that the glossy images had already been developed and were displayed atop a long table. They were available for purchase.

Of course we bought them. The next time I visited India, I took the photo with me to the village, where no one was all that impressed. Within hours, word spread among the villagers that Dr. Shukla had been seen with a man wearing a long skirt.

This was a comparatively benign rumor. Once, when someone learned the price of a plane ticket from the United States to India, the gossip was that I had purchased an airplane. When the national news reported that I had helped India establish the first supercomputer center for weather forecasting, word spread in the village that "the boy who grew up in the soils of this village has discovered supercomputers."

A BILLION BUTTERFLIES AT LAST (AND A BILLION FISHES TOO!)

By the mid-1990s, the ideas and concepts I had championed were making life better for people around the world, but my own personal scholarship on seasonal predictability hadn't enjoyed the same success. The paradigm of the butterfly effect remained deeply entrenched in the scientific community—and society at large. Thanks to Lorenz's 1972 lecture "Does the Flap of a Butterfly's Wings in Brazil Set Off a Tornado in Texas?"; *Chaos*, James Gleick's 1987 bestselling book; and *The Butterfly Effect*, a 2004 film starring Ashton Kutcher, the butterfly effect has become a universal symbol for chaos and a universally accepted idea about the tyranny of chance in our lives.

I spent my career wrestling with butterflies in the context of seasonal prediction making great progress by discovering the influence of boundary conditions, the ones I spoke about in Reading in front of Edward Lorenz.

That talk was based on a two-part paper I had submitted to the *Journal of Atmospheric Sciences* in 1981. Part one showed that monthly averages should be predictable thanks to a specific property of the planetary scale waves in the atmosphere—they change slowly and dominate the variability of seasonal averages. Part two of the paper showed that boundary conditions like sea-surface temperatures have far more influence on monthly and seasonal averages than initial conditions do and therefore provide a much stronger basis for predictability of monthly and seasonal average temperatures and rainfall.

The journal accepted part one of my paper for publication but rejected part two. The reviewer alleged that I had merely given examples of specific years using sea-surface temperature and had not established a firm basis for predictability of monthly and seasonal averages. How could I say that the butterfly effect would not eventually produce enough noise to overwhelm the boundary-forced signal? the reviewer asked. Ultimately, there was so much new evidence emerging from the research by COLA and other scientists worldwide that I decided to drop the idea of publishing the paper.

However, the question the reviewer asked continued to gnaw at me. I needed to find a way to demonstrate convincingly, in Lorenz's language of predictability, that the butterfly effect could not overwhelm the influence of strong boundary conditions, and I had to design a numerical experiment that would prove it for a complex model of the global atmosphere. Finally, after many years, a way to do that came to me—a billion-butterflies experiment. Rather than changing the initial conditions of a model only slightly, like altering the flaps of a single butterfly, and running the model for a season, what if I changed the initial conditions drastically—like altering the flaps of *billions* of butterflies—and ran the model again using the same sea-surface temperatures? Then I could see whether those flapping wings overwhelmed the effects of the ocean-boundary conditions or if the ocean-boundary conditions overwhelmed the butterflies.

After examining the weather maps and recorded sea-surface temperatures over many decades, I decided that the initial conditions of December 1982 and December 1988 provided a good contrast on which to test my hypothesis. During these two years, atmospheric and oceanic conditions were nearly opposite of each other: 1983 was a year of El Niño and 1989 was a year of La Niña.

I integrated the global atmospheric model with these two very different initial conditions, first December 15, 1982, and then December 15, 1988.

When I looked at the results, I had trouble believing what I was seeing. Those two wildly different sets of initial conditions somehow produced essentially identical average rainfall the following springs. The two forecasts had begun to *converge* rather than diverge, and within two weeks they were essentially the same.

I showed these results to some of my students and colleagues. Some suspected that there had been a computer programming error; others joked it was some kind of magic. This was not a simple toy model; it was a state-of-the-art, complex global atmospheric model, and I had begun the calculations with global observed initial conditions. Such models were not known to converge to identical solutions, especially when the initial conditions were as different as possible in nature. We checked the results again. There were no errors. There was nothing wrong in the calculation, and there was no magic, just two completely different atmospheric conditions converging if the ocean conditions were the same. When I repeated the experiment with another ocean temperature but two different initial atmospheric conditions, the two solutions once again converged.

These experiments had at last proved that the effects of sea-surface temperature were so large that *billions of butterflies* in the initial conditions could not change the outcomes—a strong and robust scientific basis for dynamical seasonal prediction.

Another unexpected result of great practical value I found was that in those years when there was a large tropical influence of sea-surface temperature, the seasonal averages in both the tropics *and* the extra-tropics were more predictable.

Since the two solutions for tropical rainfall and extratropical circulation had converged very close to what was observed during that year, this experiment showed that the models had reached such a high level of fidelity that if the correct values for sea-surface temperatures were provided, they could correctly simulate the atmospheric circulation and rainfall. This had huge implications for dynamical seasonal prediction; if we could build realistic coupled ocean-atmosphere models that correctly predicted tropical SST conditions, we would be able to make dynamical seasonal predictions as routinely as we made weather forecasts.

So COLA scientists embarked on ocean-predictability studies for an ocean model following a similar procedure that we had used for the predictability of the atmosphere. We ran a global ocean model with the same atmospheric conditions but two very different initial ocean conditions—following the butterfly metaphor for the atmosphere, eddies and waves caused by the fins of billions of fishes. The ocean conditions for both model runs began to converge slowly until they were very close to each other and close to the observed SST. However, unlike the tropical atmosphere, which converged in two weeks, the tropical ocean converged in three to six months, a notable finding because it gave hope for predicting El Niño.

This success of these experiments reminded me of something theoretical physicist and mathematician Freeman Dyson once said: "Science originated from the fusion of two old traditions, the tradition of philosophical thinking . . . and the tradition of skilled crafts. . . . Philosophy supplied the concepts for science, and skilled crafts supplied the tools." When I look back at my own audacious conjecture that, in spite of the butterfly effect, some aspects of weather and climate must be predictable, I can only think of my general optimism about life and my philosophical thinking that nature cannot be so cruel to such a vast segment of humanity. I just had to figure out how to use the tools to prove it.

All of this was the culmination of many years of effort and, frankly, optimism, hope, and devotion, but it was also a strong justification for pursuing the goal of routine dynamical seasonal prediction using complex global coupled ocean-atmosphere models.

Unfortunately, those coupled models are not being developed as quickly as they should be. There are only so many supercomputer centers in the world, each with a limited number of staff and financial resources, and the majority of them are dedicated to proving—again and again—that humans are affecting the Earth's climate, thanks to society's inability to accept this fundamental truth. There are not enough scientific and computational resources left for research on building reliable coupled ocean-atmosphere models to make accurate and reliable routine seasonal predictions, which are critically needed by society now and for a climate-changed future.

Regardless, after these experiments, I felt satisfied that I had done what I had set out to do—provide a convincing and unambiguous demonstration for predictability despite the butterfly effect. By that point, the scientific basis of seasonal prediction was so widely accepted that I did not feel the need to publish the results of these experiments. But a few years later, after I gave a lecture about this work at a conference, a young woman approached me and asked if I had submitted these results for publication anywhere. It happened that she was an editor at *Science*, and eventually, that was where my article "Predictability in the Midst of Chaos" was published.

Thirteen

My COLA colleagues and I might have been on the cutting edge of our field, traveling around the world to meet with prime ministers, presidents, and popes, but even we were not immune to frustrating office politics. By the early nineties, it was becoming increasingly clear that many at the University of Maryland weren't too thrilled to have the center in their midst, despite our headline-making work that reflected well on the institution and attracted more students to the department than ever before.

Some of their annoyance, admittedly, was my own doing. In my very first meeting with the dean, he asked me to estimate the current standing of the department of meteorology, and I answered him candidly; I told him it was probably somewhere in the middle 33 percent of programs in the United States. At this, the dean looked puzzled. "But the department claims they are in the top ten percent!" he said.

I had to explain to the dean that our research team had not come to Maryland because of its academic excellence but because it was convenient; it offered us tenured positions and room to grow. I promised him that getting the department into the top 10 percent of meteorology programs in the country was among my most pressing goals.

I was only trying to be helpful, but it seems my words did not stay between the dean and me. Things started going south pretty fast. It was a lot of little things, like when a few faculty members protested a weekly lunchtime seminar we held because the space happened to be where they played cards. And bigger things too; tenured faculty asking our research scientists to leave departmental meetings and wait outside during certain discussions, as if these dedicated university employees were foreign spies. We were overlooked for important leadership roles; we were accused of disrupting the dynamic that had been in place when we arrived; we were even threatened with expulsion from the departmental office space. All of this took a toll, distracting us from our essential work.

The last thing anyone expected us to do was leave the school. Centers like ours depend on universities for more than legitimacy; the university provides the office space, houses the computers, and administers payroll and benefits packages. Our peers and funders—the National Science Foundation, NOAA, and NASA—were comfortable giving us two or three million dollars a year because the University of Maryland took care of administrative details while we worked on the science their dollars funded. Furthermore, at the University of Maryland I had tenure—a lifelong guarantee of employment. It's not a thing people want to walk away from.

The other barrier to leaving was the impracticality of moving a team of people and their families to another institution. It's hard enough for married people to get joint appointments at the same school; it would be difficult to find a university willing to hire the dozen or so researchers we employed at the time.

I guess I should have felt stuck, but I didn't. There were significant accomplishments of the past decade under COLA's belt—seasonal prediction, reanalysis, the simple biosphere model, our contributions to national and international research programs, and the impact of our science globally. When a scientist from Japan visited COLA, housed in part of one floor of a commercial building, he asked if COLA occupied the entire building. I felt strongly that the federal agencies we relied on would continue supporting our work no matter where we were. Of course, *where* was the million-dollar—actually, the two-million-dollar—question. What if we made our own *where*? I thought. What if we just started our own institution? I had had a lifetime of crazy thoughts, but this was perhaps the craziest yet.

Still, when I sat down with Jim, Ed, and David and told them what I was thinking, they didn't act like my idea was crazy. Jim and David were fully on board; both of them were ready to resign. They endorsed the idea of our own center enthusiastically. David's approval was especially buoying to me. He had already resigned from a high-ranking permanent appointment at NASA to join the center; now he was ready to take another giant leap of faith. Ultimately, we knew, each of us could find jobs if things didn't work out and the funding didn't come through. But the goal was to stay together and continue the work we were doing—this time, we hoped, without the squabbles over space and seats at the conference table.

And so, during the day, we did what we always did—taught classes, worked on research papers, mentored Ph.D. students. But at night and on weekends, we stayed up late reading books about establishing an independent center, talking to other leaders of nonprofits and lawyers, strategizing over containers of takeout food. We explained to our contacts at NASA, NOAA, and NSF that although we were working together as a team on a focused single project of dynamical seasonal prediction, they had been funding us as if we were working on ten separate projects. The three agencies agreed to accept a single proposal (instead of the ten we had been submitting for years), saving us untold hours of paperwork. However, the agencies warned us that we were taking a massive risk by submitting a single proposal, since it would take only one rejection from any of the three agencies to end the new center before it started. We were willing to take the risk.

The first action we needed to take was to officially form a nonprofit organization, an umbrella corporation that would apply for and administer grant money to COLA and provide the kind of logistical support that the University of Maryland had been giving us for a decade. The Geophysical Fluid Dynamics Laboratory has Princeton; the National Center for Atmospheric Research has the University Corporation for Atmospheric Research; NASA's Jet Propulsion Laboratory has the California Institute of Technology. We decided that COLA's nonprofit would be called the Institute of Global Environment and Society, IGES. And the founding employee we chose to manage federal grants was none other than Anne.

Anne knew a lot about how to manage a center like ours, thanks to the work she had done for the Architecture Machine Group at MIT. An MBA, Anne was well equipped to apply for all the right licensing, set up payroll functions, investigate benefit packages, and keep meticulous records that could stand up to rigorous federal audits. She took the job on enthusiastically, but it was tedious work and a lot to task her with, especially since she was doing the majority of the child-rearing on her own. One benefit she enjoyed was the ability to work from home. And that was because, for the first six months, the Institute of Global Environment and Society was located in our home. In our garage actually.

The hardest problem was setting up an independent, federally funded research center without any money to start with. The National Science Foundation could not offer an advance, but administrators assured us they would reimburse salaries and other costs once we were awarded a grant. But in order for them to give us that grant, we needed to demonstrate that we had the space, equipment, and infrastructure to run a successful organization. All of this required setting up a significant line of credit, and because we were a brand-new organization, *that* required significant collateral.

I was so boldly confident that using Anne and my home for collateral didn't strike me as all that risky. But when I told friends and colleagues about the decision, they used words like *stupid* and *crazy*.

Even my contact at NSF, an old friend from MONEX who had been largely supportive of our efforts, warned me to be more cautious. I didn't listen.

That's how our suburban DC home became both collateral and headquarters where we would work until we had the money to rent a proper space. Conveniently, our house had been the model home when the neighborhood was built; its garage had been the sales office with sixteen telephone lines. It was finished and large enough to squeeze in a handful of desks and a smattering of computers. Once everything was in and set up, I thought the place looked halfway professional.

All of this work happened over the course of six months or so, and the last hurdle was a site visit from the management of the National Science Foundation. It was the only time I remember feeling nervous. We had put everything on the line—our jobs, our home, the future of COLA itself—and now we would find out if we would actually get the money we needed to carry on. The person deciding our fate was a serious middle-aged woman in a pantsuit, and when she looked at the desks in the garage, unlike me, she didn't seem all that impressed. "Where is IGES?" she asked.

After a few awkward moments, the woman took a seat at our kitchen table. Jim, Anne, and I joined her. (Upstairs, Chandran and Sonia, now twelve and ten years old, were reading quietly; Anne had put the fear of God in them about staying out of the kitchen, telling them it was strictly off-limits that afternoon.) The woman looked at me, then at Jim, then at Anne, then back at me.

"What happens if you die?" she said abruptly.

Taken aback by such a grim opening question, I stuttered a bit when I answered. Luckily, my brilliant wife began explaining the succession plans she had drawn up.

After about an hour, the NSF representative, whose scowling face is permanently imprinted in my memory, had no further questions. By the end of our meeting, she understood our mission and gave us a number of helpful suggestions.

The days after the visit, as we waited to hear our fate, were some of the most stressful of my life. Not since the panicky hours following my father's fall had I felt my heart pound so hard in my chest. It was as if my body suddenly understood what was at stake; I felt like a car growling while idle, ready to move forward but stuck in position.

Sleep was elusive. One night, I couldn't fall asleep at all. I started thinking about the loans we had taken out, about the many families that depended on this NSF money coming through, about the college tuition our children might never see if the answer was no. I began pacing up and down the stairs, disturbing Anne and making her wonder if her husband had finally lost his mind.

The sun was barely an orange smudge on the horizon when the phone rang. It was the deputy director of the International Centre for Theoretical Physics in Trieste, Italy. He was calling to tell me that although I had turned down the position at ICTP, there was still one hundred thousand euros that Dr. Salam had asked them to set aside as a salary for me in case I changed my mind. I couldn't believe the timing. It was enough to ease my panic and allow me to rest for a few hours.

Days later, we got the letter saying COLA would receive five years of funding. Now we could focus on our research in the best possible environment. Together, all the COLA scientists submitted their resignations to the University of Maryland. Six months later, in the autumn of 1993, the Institute of Global Environment and Society's office was moved out of my garage and into an office park on the outskirts of Washington, DC.

To celebrate, we hosted an open house there and asked Suki Manabe to give the very first lecture in our new and improved COLA lecture series. We invited everyone—former colleagues from Maryland and staffers from NOAA, NASA, NSF, and all the other research organizations in and around the capital. I was gratified to see how impressed everyone was as they admired our new library and conference room, all the furniture we had bought using our family home as collateral. One of my old NASA friends told me that if he worked in such a nice space, his productivity would double.

Still, it wasn't really the space or the stuff that mattered; it was the family of scientists (and after all our years together, we did feel like family) who had made it happen. Founding our own research institution was a remarkable achievement. No one had ever pulled off what we had accomplished—a not-for-profit independent institution for climate research. But none of us had done it for the sake of glory or bragging rights; we had done it because we truly believed we were advancing science and that our science was making the world a better, safer place for all people.

Many mornings on my commute to COLA, I found myself wondering if I was the luckiest man in the world.

Of course, science is a long and plodding exercise. And while the experiments we did and papers we wrote often made big splashes in the world of meteorology, the real influence of our work took decades to unfold. Now, more than thirty years after our little center made a big name for itself, I can see our true impact—and I can say for sure that I *was* the luckiest man in the world.

During the final decade of the twentieth century and the first years of the new one, thanks to our understanding of the impact of boundary conditions on climate and the rise of coupled models, dynamical seasonal prediction began to step out of the research lab and into the real world.

One of the most influential organizations deploying seasonal prediction to protect human lives and livelihoods is the International Research Institute for Climate and Society at Columbia University (IRI). I was a member of the committee that wrote the proposal for the institute. My old friend Antonio Moura was the first director general of IRI, which was founded in the early 1990s to prepare El Niño forecasts. Since then, IRI's mission has expanded; the institute now works to provide developing countries with the seasonal forecasts they need to, among other things, ensure agriculture production, anticipate migration patterns, and prepare for natural disasters and disease outbreaks.

This work is remarkably tailored to each nation. For example, IRI researchers identified the average onset dates of the rainy season across seven different agricultural zones in Vietnam. That means the farmers there now have a better idea of when to plant based on their location in a country that is smaller than California.[1]

Other countries benefit from IRI's forecasts by preparing for disasters ahead of time instead of reacting to them after the fact. In 2008, scientists at IRI predicted a high risk of above-normal rainfall for the nations of West Africa, a low-income region where the majority of the population reside in vulnerable shantytowns built on floodplains. Equipped with this knowledge, local offices of the Red Cross and Red Crescent requested funds in advance of the anticipated flooding, set up disaster-relief stockpiles across multiple countries, updated their contingency plans, and gave a heads-up to community officials in the region. When the predicted heavy rains arrived, this science-informed preparation resulted in a marked decrease in the number of lives and property lost compared to the previous year's flooding, when three hundred people died.[2]

More recently, IRI developed a system to forecast the hot, moist conditions across Central America that are especially suitable for mosquitoes and therefore the outbreak of vector-borne diseases like zika, dengue fever, and chikungunya. As the climate continues to change, the burgeoning field of climate epidemiology will put seasonal prediction to good use.

These are only a handful of ways dynamical seasonal prediction is changing the world, and today IRI is just one organization doing this work. Here in the United States, people have grown so accustomed to receiving seasonal predictions, they might not even notice it anymore. The National Weather Service posts three-month outlooks for temperature and precipitation.[3] NOAA provides a drought outlook over the same time spans.[4] Each year before the hurricane season, coastal officials and residents await with keen interest that year's hurricane-season outlook. During much of the time I spent writing this book, the nations of the world were preparing their electrical grids, agricultural systems, emergency response plans, and food supplies for a major El Niño.

And in rural India, in the villages that look like the one I grew up in, seasonal prediction has finally arrived. The India Meteorological Department has at last invested in complex coupled models and now provides long-range forecasts to the country's Ministry of Agriculture and Farmers' Welfare. Whereas my neighbors once relied on omens and waited for a glimpse at the *panchang*, they now rely on science and wait for the ministry's extension officers to relay the model's predictions and give advice about when and what to plant. It is not perfect, and there is much work to be done, but it is a vast improvement over the strife and struggle I witnessed in my youth.

If you aren't a farmer, a minister of agriculture, a humanitarian aid worker, or the owner of an expensive beach house, you might easily overlook the role seasonal prediction plays in your life. But that is a good thing. Seasonal prediction has become sufficiently reliable and routine that it is notable only when something goes very wrong. It is my hope that seasonal prediction soon disappears into the background of everyone's life, especially the most vulnerable among us.

FLASH, BOOM; FLASH, BOOM

Authors often use thunderstorms as a symbol for chaos or a plot device to foreshadow sudden and shocking change; think of King Lear's descent into madness during a tempestuous act 3 or lovelorn Heathcliff departing Wuthering Heights in the midst of a violent gale. In literature and in real life, storms represent episodes of unpredictable and unpreventable pain.

People often check the forecast to decide what to wear, whether to bring an umbrella, whether to hold their celebration indoors or outdoors, or to see if there is a storm rolling in. We've talked about how the energy emitted by the sun and Earth and prevailing wind patterns play a large part in determining the temperature—but what about precipitation? What causes the storms that have everyone clamoring for their weather apps?

When thunderstorms roared across the rice fields that surrounded the village, my family would gather in my mother's bedroom, where a concrete roof offered watertight protection and muffled the growling wind. There we would huddle together with blankets or rugs, a kerosene lamp aglow in the corner. Today I look back on these memories with a kind of rosy nostalgia—all of us there together—but there on one occasion, I remember being quite scared.

I was still in primary school. The monsoon season had been exceptionally stormy that year, so much so that portions of our thatched roof had begun to fray, and the floor of our house was pocked with puddles of mud. One evening, a massive black storm approached from the east. As the peepul trees bowed in the wind, my mother called us children inside and into her bedroom, where rugs were already laid, a lamp already burning.

That night it rained so hard it sounded like someone was pouring buckets of water on our house. Every few minutes, a terrible peal of thunder seemed to crack the sky in two. I was terrified—that is, until I began to feel curious.

I noticed that before each boom of thunder, there was a flash of lightning. Flash, boom; flash, boom. Sometimes the boom followed the flash immediately; sometimes the boom lagged by long seconds. Now my brain was too busy to be scared. Why did the flash always come before the boom? I wondered. And, come to think of it, why was this storm so much worse than all the others?

Thunderstorms carry with them and release a staggering amount of energy, many more times the energy of an atomic bomb. The source of that energy is the latent heat of condensation when water vapor condenses to produce rain. When you boil water in a kettle, the energy from the heat of the stove is transferred to the steam coming out of the kettle. That water vapor carries all the energy with it as it moves out of the kettle. Likewise, all the water vapor in our atmosphere is produced by energy from the sun, which causes evaporation from the ocean and other surface water.

When the moisture in the air condenses, it releases all of that energy. Thunderstorms die after a few hours because they exhaust the energy in the moist air they bring in. Hurricanes, however, last for several days because they travel, picking up a fresh supply of moisture from the ocean below as they go. Hurricanes die after they make landfall because the supply of moist air is drastically reduced, and the roughness of the land dissipates their energy.

In the monsoon region where I first experienced storms and in tropical regions in general, there is never a shortage of moist air! What is needed is a process to lift the moist air up. One obvious process is the heating of the ground by the sun, which in turn heats the overlying air. Warmer air is lighter and rises. If the moist air rises high enough so that moisture can condense, that becomes a new source of energy to lift the moist air higher and higher—until the rain begins to fall. We do not see thunderstorms all the time even though moist air is always present because, in addition to the heating of the ground, we need vertical profiles of temperature and moisture in the atmosphere so that the moist air has the buoyancy to keep going up and up; these are not always present. The physical and dynamical processes that cause the lifting of the surface air to produce storms in the tropical regions are different from the processes that produce storms in the extratropical regions. That is why our ability to predict weather differs between tropical and extratropical regions.

Okay, so that's how storms get their start, but what about the stars of the show—the thunder and lightning? In a tall thunderstorm cloud, friction among water droplets and ice crystals of different sizes and weights can create static electricity. While larger, heavier droplets remain low in the cloud, smaller and lighter droplets are carried higher and higher. The big drops tend to accumulate negative charges, while the littler drops accumulate positive charges. When these opposite charges become large enough, a violent electrical discharge takes place—lightning! Sometimes lightning traces a path from a negative charge in the cloud to a positive charge on the ground or to a positive charge in a nearby cloud.

Thunder is created by lightning, which heats the air rapidly, causing it to expand. The air in a lightning channel can reach temperatures as high as 50,000 degrees—much hotter than the surface of the sun. But only for a fraction of a second. After the flash of lightning, the air cools and contracts as quickly as it heated up, creating the sound wave that we call thunder.

It takes the sound of thunder five seconds to travel one mile. To figure out how far away a particular bolt of lightning is, simply count the seconds between the flash and boom and divide by five. Fifteen seconds, and that lightning is three miles away. Five seconds, and it's only one. Zero seconds? You'd better be inside already.

In the story of my life, I was about to find myself outside and without cover, completely oblivious to the gathering clouds.

Fourteen

After the successful prediction of the 1997/1998 El Niño season, when Ants Leetmaa cracked his joke about my imminent retirement, what he didn't know—what no one knew—was that, as I neared my sixtieth birthday, I was beginning to think about stepping away. During a COLA retreat, I had even broached the topic of future leadership, the first time I had ever mentioned it.

Dynamical seasonal prediction was happening in the real world, saving and improving real lives, and most of the major weather centers were now working on making their own long-term forecasts. I still had a strong desire to be of service to society and I knew there were much bigger problems than the ones meteorology could address. Hunger, inequality, poverty—what could I do to ease the suffering that plagued so many? It was a question I wanted to try to answer.

Around this time, I read an article in *The Washington Post* about Michael Jordan, who, like me, was preparing to retire. The headline announced that the basketball player was leaving the sport "at the top of his game." The words resonated. I was not a legendary basketball

Toga party at the Melbourne jail at the end of the Tropical Ocean Global Atmosphere program; Shukla, Ben Kirtman, Ed Schneider, circa 1995.

Ed Lorenz visiting COLA, 1996.

This flag with COLA logo, signed by all COLA scientists, was taken by astronaut Piers Sellers on his Space Station Assembly Mission (STS-112/9A) on October 7, 2002.

Piers Sellers.

Founding Climate Dynamics Faculty at GMU: Paul Schopf, Tim DelSole, David Straus, Shukla, Ed Schneider, Jim Kinter, Ben Kirtman, V. Krishnamurthy, Barry Klinger, Bohua Huang, 2003.

Antonio Divino Moura and Shukla at the inauguration of the Asia Pacific Climate Center, Busan, South Korea, 2005.

Sonia Shukla, Anne, Shukla, Ed Lorenz during the Ed Lorenz symposium, American Meteorological Society, San Diego, 2005.

Dr. Manmohan Singh, Prime Minister of India, invited Shukla at his residence, New Delhi, 2006.

Receiving the International Meteorological Organization (IMO-2007) Prize from Alexander Bedritsky, president of the World Meteorological Organization.

A group of former GMU PhD students of Shukla gather for dinner: Susan Bates, Laura Feudale, Kathy Pegion, Julia Manganello, Anjuli Bamzai, circa 2007.

Tim and Gill Palmer, Susie and Mike Wallace, Ken Mooney, Peter Webster visiting Shukla's village and Gandhi College, 2008.

Mrs. Pratibha Patil, President of India, presents Padma Shri—one of the high national awards—to Shukla at Rashtrapati Bhavan (President's House), New Delhi, 2012.

Inez Fung, Antonio Moura at Shukla symposium (and forty-two years earlier in black-and-white photo), 2015.

Shukla, Arthur Bass, Inez Fung, Robert and Nora Charney Rosenbaum, Eugenia Kalnay, Milt Halem, David Straus, Joe Pedlosky at the Charney-Lorenz symposium at MIT, 2018.

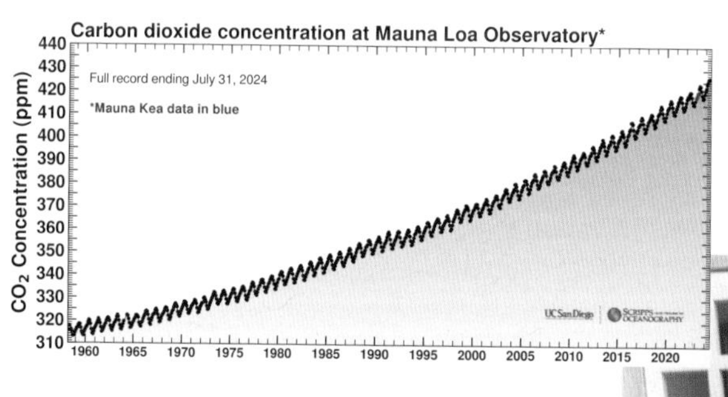

(Above) Carbon dioxide concentration at Mauna Loa, Hawaii, Observatory, which has been steadily increasing since measurements were started in the 1960s. It shows a prominent seasonal cycle reflecting change in photosynthesis and vegetation.

(Right) Suki Manabe received the 2021 Nobel Prize in Physics at the National Academy of Sciences, Washington, DC, 2021.

Shukla's family on his eightieth birthday. Sitting: Vinod and Pooja Dubey; J., Anne, and Sonia Shukla; Natasha Shukla Qazi. Standing: Aastha, Aarushi Dubey.

Participants at the Shukla symposium in April 2015, Rockville, Maryland.

player, but I did see the wisdom in calling it a day when the day was still young. I cut the headline out and tacked it above my home desk; I began drafting my retirement speech.

My first foray into a new line of work began well before it was time to give that speech, and it happened, naturally, in Mirdha. During one of my short visits—a day or two I had no doubt wedged in between flights to other far-flung places—my mother said something that caught me off guard.

"You are doing so much around the world, going here and there"—she swung her hand like a pendulum—"but what have you done for the village?"

It was a fair point. I had been working on the advancement of climate science in the United States, Italy, Brazil, Korea, and India—just about everywhere but my village. Yet each year when I went back, I would step inside the primary school my father had built, the one that *still* served as the village's only educational facility, and marvel that nothing about it had changed. There were still no chairs, no desks, no doors. Dried animal droppings and dead leaves littered the corners. When I asked one of the village leaders how many students made it to the end of their primary school years, he guessed around 30 percent.

After my mother's pointed question, I knew what I needed to do, but it hardly felt like my idea alone. Opening up a community college in the village was merely a continuation of the work my parents had started. I still remembered how much my father had to fight to get that two-room school built. And watching my mother give away our things—our rice, our blankets—to every hungry and desperate beggar who came to our door was a memory my mind could still conjure in full color and crisp resolution, despite the decades that had gone by. It took a few years to convince some skeptical villagers, and Anne and I had to make the commitment of family land and money, but Gandhi College opened in Mirdha in 1999.

The project turned out to be far more challenging than I had originally thought. The members of the village committee that I had formed to get advice and guidance were always proclaiming that corruption was the biggest problem in India; however, they did not hesitate to take part in that dysfunction themselves. The state government officials, who had the authority to approve the establishment of a new college, wanted a bribe at every step. Even the Home Ministry in New Delhi, which had to approve the transfer of funds from the United States to the new college, wanted a bribe. I was faced with a real dilemma: Should I start Gandhi College with a bribe? I was told that it simply was not possible to create a new institution without bribes. One person was quite blunt. "Do you think that major humanitarian organizations are doing all these big projects without any bribes?" he asked. But early in the process, I made a firm decision that Gandhi College would not offer bribes, and I intended to stick with it.

Instead, I worked with some officials in New Delhi. My position as director of COLA in the United States and my work on the supercomputer and monsoon forecasting in India helped me get meetings with important people who normally wouldn't care about what happened in a village like mine. Still, it took many months to get the approvals. In the village committee's view, I wasted a lot of time; bribes would have done the job much faster. But I wanted the villagers to know that honesty and transparency would be a cornerstone of the college.

Gandhi College could not afford to provide a science and engineering education, but it could strive to build character and empower girls. Based on the Gandhian principles of honesty, perseverance, and selflessness, the college is a coed school, but 80 percent of its students are female. This was by design; until there was a local option, most girls in the village were denied post-secondary education. Either their parents couldn't afford it or they prioritized their sons' educations with the little money they did have. Many young women simply weren't allowed to live in Ballia, more than half a day's walk from home.

My younger brother Shriram, who had a green card and could have stayed in the United States but had returned to India, worked very hard to supervise the construction of classrooms, laboratories, and a library and to get the academic program of the college established. He has overseen the college for the past two decades. In that time, we have witnessed the self-esteem of women from our and neighboring villages soar. As the son of a mother who couldn't read or write, I find great satisfaction in witnessing a new generation of women who are educated, empowered, and know full well the respect they are owed. Perhaps best of all, I had satisfied my mother's appeal. It turns out that making her proud felt just as good when I was sixty as it had when I was six and lighting candles in the prayer room.

A year or so after the college was up and running, my mother's health began deteriorating, and in April 2000, she passed away. Attending to the rituals surrounding her funeral required me to be in the village for three weeks, by far the longest I had spent there since I left for MIT, but there was nothing that could keep me away from saying goodbye.

In Hindu mythology, the soul never dies. Death simply means that the soul has left one particular body, and after some time, it will go on to inhabit another body. This is the basic premise of reincarnation. The purpose of many Hindu funeral rituals is to ensure the soul's peace and safe passage from one body to another. These rituals—which include daily prayers for the peace and tranquility of the departed soul and offerings of water and food—can last for up to two weeks. For a Brahmin like my mother, it lasted for thirteen days.

On the ninth day of the funeral rituals, the other men in the family and I shaved our heads and faces in a show of mourning. On the thirteenth day, we hosted a massive meal and held an elaborate prayer to make certain my mother's soul was finally ready to move on to its next destination.

It was a poignant, languid time and I was grateful for the chance to honor my mother's life in such a deliberate manner, especially after my father's sudden and unexpected death. But during the many days of rituals, I was also reminded of the volatility of my culture and I was plunged back into the familiar chaos that had defined my childhood.

For instance, the priest that performed the funeral rites suggested that lots of cash, cows, and gold be offered—lest the soul feel restless and lost. My brothers, eager to please the priest and guarantee my mother's successful reincarnation, did whatever he suggested. When my two brothers, accompanied by about two hundred and fifty relatives and villagers, carried my mother's body to the Ganges for cremation, another priest, a special kind that provides fire to light the funeral pyre, realized that our family was well-to-do and asked for fifty thousand rupees to give the initial flame. Luckily, in this instance my brothers showed some restraint and negotiated him down to five thousand.

As for the big dinner on the thirteenth day, Mahendra spared me no expense. Knowing I would happily foot the bill for the feast, he sent messengers to all the surrounding villages with drums to beat and announcements to make. My best estimate was that eight thousand people showed up.

Resigned to the madness, I sat on my father's old sleeping porch and watched as the cooks arrived and dug ten holes side by side in which to build their fires. In improbably large cauldrons, about fifty cooks stewed a number of vegetables. They dropped dough into pots of hot oil to make *puris*, stirred up vats of chutney, and rolled out trays of sweets. About seven hundred people could eat at a time, and meals lasted about twenty minutes. During each seating, dozens of laborers circulated around the open field serving the attendees, who ate their food from plates made of leaves and drank from earthen cups they tossed into piles when they were finished. (I had been to events like this before, but watching thousands of people *sustainably* having dinner was a new experience for me.)

The smells, the yelling, and the frantic pace resembled a circus more than a funeral celebration; it was more entertaining than any reality TV competition you can imagine. Around midnight, I finally grew tired of the spectacle and went to bed. When I woke up in the morning, there were enough leftovers to feed about two thousand people, and the food was distributed among the villagers.

Before my trips to India, people always ask me what it's like to wake up in Washington, DC, a city stuffed with ambition and expensive suits, and go to bed in Mirdha, a rural enclave where residents still use the fields as their bathroom. I tell them I have three lives, each as important and visceral as the others. One in America, one in urban India—places like Delhi and Pune—and one in my village. As I journey from one place to the next, I am able to seamlessly fold myself into them without thinking much about it, speaking the local dialect, wearing the traditional clothing, performing the customs with the ease of a native. My friends in Maryland have a hard time picturing me chewing betel leaves and chattering with passersby in Bhojpuri, but it is as natural to me as sitting in a conference room sipping on bottled water.

When I traveled to my mother's funeral, I cycled through those lives just as I had done countless times. But when I left India afterward, I couldn't help but feel that a big part of the third one—the life in my beloved village—was coming to a kind of end. To me, my mother was the embodiment of the village, my strongest tie to Mirdha. When I thought of home, I thought of her. When I thought of her, I thought of home.

Moreover, some of my family members had immigrated to the United States, creating something like a village in the middle of Maryland. Most important, Pooja had come to live with us and attend college in America. It wasn't the easiest transition—everyone in our family of four had to reconfigure our roles as we welcomed a fifth member—but having my entire family under one roof was a long-awaited relief. Sometimes having so many lives is great fun, full of adventure; sometimes it is painful, rife with longing.

Perhaps in the past I could have tried to convince my mother to come too, but I know that would have been a losing battle. She came to visit the United States only once, during the frenzied few years when the children were very young. Anne and I did our best to make her as comfortable and happy as possible (including ridding our entire kitchen of any evidence of meat and fish), but I'm afraid my mother was quite unimpressed by my adopted country. In fact, she was bored.

She spent a lot of time perched in front of a large picture window watching cars and people go by. She couldn't understand it, she told me—people passed each other on the sidewalk without so much as a glance. In the village, talking, gossiping, and visiting were not mere distractions—they were the essential business of everyday life. When it was time for her to go home to Mirdha, she was more than ready.

It would be an understatement to say my mother was proud of me. When she got back to India, she began exaggerating about my status in America. She told my former neighbors I wasn't just a professor at the University of Maryland—I was its president.

It was on this visit that she told us Chandran reminded her of Mahendra. He was so young then, just a toddler, and I wasn't sure what she meant. But as my son grew, it became obvious that my mother was right and that Chandran had inherited the daring, adventurous, and independent streak of my older brother.

Growing up, Chandran was often bored in school, which always moved too slowly and quietly for his preference. Endowed with an inexhaustible curiosity and nimble hands—not to mention an indomitable sense of confidence—our son eschewed books for action, whether he was building a computer, launching rockets, or tearing apart car engines. He gravitated toward loud, combustible things, even guns. Once, while he was attending college in Utah, Chandran was pulled over twice in one day, both times for driving more than one hundred miles per hour. This was just one of many moments that made his parents—who enjoyed afternoons spent with the newspaper and evenings having dinner with friends—turn to each other and wonder just where this boy of theirs had come from.

Just as Mahendra had defied my father's wishes by insisting on becoming a wrestler, Chandran was utterly uninterested in the plans that Anne and I had for him. But it was hard to say no to Chandran. His smile was enormous, his eyes as bright and insistent as wildfire. He was also smart enough to win just about any argument he started. And that is why I have been to not one but two gun shows and several drag races. I knew that if I wanted to spend time with Chandran—and I very much did—I needed to encourage his interests, no matter how far they diverged from mine.

The one instance when Anne and I drew the line was with the motorcycle. When Chandran asked us to help him buy one, we said absolutely not. We pleaded with him to reconsider the idea.

Nevertheless, one morning I was awoken in the gray light of dawn by the beeping of a truck backing out of the driveway. As I walked down the hall to investigate, I glanced in Chandran's room—he wasn't there. He was already downstairs, where he had just signed the bill of lading for his new motorcycle. Within weeks, he had joined a racing club. He promised us that he would be careful.

We were much more supportive of his other gutsy ambition: flying planes. Chandran wanted to be a fighter pilot. When he came home from his months in Marine Corps boot camp, it was as if he had been remade; his arms were steely and his manners impeccable. He said "Yes, sir," and "Thank you, ma'am," to the guests at the big party we threw for him. Anne and I were so proud, so ebullient to see our son channel all his intelligence and energy into an endeavor that would make the world a better place. We couldn't wait to see all that he would accomplish. But that was not to be.

On March 14, 2004, Chandran was riding his motorcycle with friends on Skyline Drive, the twisting ribbon of road that traces the ancient Appalachian ridges of Shenandoah National Park, when his wheels hit a patch of loose gravel. Our son was not yet twenty-three and he was gone.

It is not possible to describe the immensity and depth of our grief. It was confusing and improbable—a startling reminder that we were in control of nothing. In the days and weeks after Chandran's death, I struggled to make sense of the randomness and the arbitrary nature of tragedy, the stealth and surprise with which it strikes.

For months, Anne and I were lost. In the middle of a mundane chore, cooking dinner or folding clothes, we would begin to cry uncontrollably. Being around or talking to Sonia and other family was the only antidote for the crying, it seemed. In fact, for a long time, the only thing Anne and I felt like we could do was batten down the hatches and draw our family close, the way my mother had during the monsoon storms of my childhood.

And when it was time, somehow, for life to go on, I discovered that the whim to retire from meteorology had all but vanished. I felt quite the opposite, actually; I wanted nothing more than to sit at my desk and work on the same science that had sustained me for years, through my father's death, through my emigration and the painful decision never to go home for good, through the upheaval in Maryland, and, now, through the loss of my son. The idea of diving into something new or different no longer felt alluring or exciting. While many in our extended families sought comfort in their respective religions, I found it in the familiar rhythms of work and research. Instead of slowly winding down my career, I began to say yes to every project and proposal that came my way.

So when Peter Stearns, a great scholar and the provost of George Mason University, where COLA scientists had been teaching courses for years, told me the university was interested in forming new Ph.D. programs, I told him I could put one together in less than a year.

It was ambitious, but the busyness was soothing. Over the next few months, I designed a whole new department for GMU, the department of atmospheric, oceanic, and earth sciences. In our vision, it would provide the holistic approach to climate that COLA had pursued for decades but that would be totally novel in a university setting. Students could earn a doctorate in climate dynamics, as I named the program, and they could train on COLA computers using one of the most comprehensive data sets and advanced coupled models under the tutelage of some of the best climate scientists in the world. The department would be created by combining the ocean, land, and atmosphere faculty of COLA and the geology faculty from another department.

All we needed, I told the provost, were eight faculty positions, and I would hand over the reins of COLA to George Mason; all of the research grants from the federal agencies that funded us would go to them. All the accolades from our innovative science would be theirs to parlay into more money, more students, more prestige. I didn't actually think GMU would agree, but not only did the university provide the funding for eight new faculty positions, but the dean and the provost constantly encouraged and supported us. A new department of atmospheric, oceanic, and earth sciences was born. To our great delight, three years after COLA scientists and research projects transitioned to the university, GMU was classified as one of the elite group of the highest ranked (R1) research universities in the country.

GMU had its motivations; I had mine. After Chandran's death I realized how uncertain the future could be. I wanted to take care of the scientists who had—multiple times—taken huge risks for the sake of our scientific mission and our center. With COLA back under the umbrella of a university, many of those scientists would have a more certain future as tenured professors. In the mid-2000s, around the time I had thought I'd be retiring, we embarked on a brand-new chapter of our humble center.

I was also undergoing my own transformation. Before Chandran died, I was always resisting the tide, badgering colleagues or administrators to change this or stop doing that. But afterward, I became less like the rock in the middle of rushing river and more like the river itself. I no longer wanted to push against the current; I let it take me where it would.

PART IV

A Reason for Hope

Fifteen

Around the turn of the nineteenth century, when my father was a young boy spending his childhood much the way I would one day spend mine—traipsing around grassy fields with a tribe of cousins, receiving lessons beneath the spidery arms of the banyan tree, searching the sky for dust storms and monsoon rains—the world far outside the village underwent an incredible transformation. Combustion engines roared to life, and coal-burning locomotives charged across entire continents. Men in dirty overalls demolished vast tracts of forest and excavated minerals to fuel factories so big, they blotted out the horizon. No one knew—not the villagers, certainly, and not even the men at the helm of this furious revolution—that humanity had embarked on what scientists would later call a "large-scale geophysical experiment that could not have happened in the past nor be reproduced in the future."

That's because, just as the physical landscape in industrializing countries was undergoing a dramatic makeover (from landscape to skyscrapers, from green to gray), the makeup of the atmosphere was changing too—staggering amounts of carbon dioxide flooded the air. It would take another hundred years for people to recognize one of the greatest dilemmas of humanity: the destruction of the natural environment and the uncontrolled release of carbon dioxide that accelerated the growth and development of civilization could someday make the planet uninhabitable.

Prior to the Industrial Revolution, a handful of scientists had investigated the connection between the amount of CO_2 in the atmosphere and the surface temperature of our planet. In 1897, a Swedish scientist named Svante Arrhenius suggested that a decrease in CO_2 could trigger an ice age and a doubling of CO_2 could increase the Earth's temperature by 4–5 degrees Celsius. He did not consider that alarming, nor did his audience, because based on CO_2 emissions at that time, Arrhenius estimated that it would take about five hundred years to double the amount of the gas in the atmosphere.

In the late 1930s, Guy Stewart Callendar, an obsessive amateur meteorologist, became convinced that the burning of fossil fuels was in fact ratcheting up the heat here on Earth. To come to this conclusion, Callendar had collected historical records from 147 weather stations around the world, and after doing all of his calculations by hand, he discovered that the average global temperature had risen approximately .3 degrees Celsius in fifty years. In the same period, Callendar estimated that humanity had burned enough fossil fuels to send about 150 billion metric tons of CO_2 into the atmosphere. The two facts, he asserted in a 1938 paper, were certainly connected.[1]

This was well before the advent of coupled ocean-atmosphere models, which would provide a scientific explanation for the relationship between greenhouse gases and average global temperature. It was also before anyone was measuring CO_2 all that carefully. As there was no reliable data or solid evidence, many scientists dismissed Callendar's findings as a coincidence, and Callendar himself wasn't too concerned about what he had discovered; he believed a warmer planet would prevent "deadly glaciers" from returning and facilitate agriculture in more parts of the world.

Nevertheless, the Callendar effect, as anthropogenic global warming was first known, caught the attention of a number of prominent researchers, including Roger Revelle and Hans Suess, scientists at the Scripps Institution of Oceanography. In a 1957 paper, Revelle and Suess expressed skepticism that so much carbon dioxide was simply lingering in the atmosphere; they postulated that the average CO_2 molecule spent about ten years airborne before it dissolved into the ocean. Therefore, most of the carbon dioxide released by burning fossil fuels, according to the scientists, had already been sequestered by the sea.[2] Still, Revelle and Suess expressed caution about what they famously deemed that "large-scale geophysical experiment" of burning unprecedented amounts of fossil fuels. (That year, Revelle even testified to Congress that fossil-fueled climate change could transform Southern California into deserts.[3]) The scientists urged further study, especially as a part of the International Geophysical Year (IGY), a yearlong collaborative effort among sixty-seven countries to investigate various fields of earth science, including gravity, cosmic rays, precision mapping, seismology, and oceanography.

Revelle got his wish, using IGY funding to establish the Scripps CO_2 Program, which endeavored to measure carbon dioxide levels around the world as accurately and consistently as possible. For this, he hired Charles David Keeling, a researcher who had developed hypersensitive CO_2 monitors and had already used them to discover that concentrations of the gas were highest at night and lowest in the afternoon, thanks in large part to daytime photosynthesis. Armed with his ultraprecise sensors, Keeling established research bases in Antarctica and on Mauna Loa, a volcano on the island of Hawaii.

And thus one of the most iconic data sets in all of climate science was born. The Keeling curve, as it's known, depicts the steady upward trend of carbon dioxide in the atmosphere over the past sixty-five years. In 1958, when Keeling began his measurements, carbon dioxide made up 313 parts per million in our atmosphere (this means if you took one million molecules of air, 313 of them would be CO_2 molecules). He discovered that, just as CO_2 concentration fluctuates during a day, it also fluctuates seasonally; CO_2 is highest during May and lowest during October. As leaves and other plants on the ground decompose during winter and microbial respiration produces CO_2, the CO_2 steadily increases in the atmosphere; it reaches its peak value in May and then begins to decline as plants and trees take in more of the gas as they grow during spring and summer. But as time wore on, despite this "seasonal breathing" of plants, which created a sawtooth pattern in the graph, it became apparent to Keeling that carbon dioxide present in the atmosphere was increasing, year after year. By 1965, CO_2 made up 320 ppm (as of this writing, it's close to 425). There was no dearth of skeptics and detractors when he began taking measurements at Mauna Loa. Some wondered if the increases in the early period were due to the large number of tourists driving their cars there and idling their engines near the observatory.

The Keeling curve was the first depiction of the stark and undeniable surge in CO_2 in the Earth's atmosphere, a phenomenon scientists were fairly certain would have a significant impact on global temperatures. During the late 1960s and early 1970s, an era that saw the publication of Rachel Carson's *Silent Spring*, the first Earth Day, and the formation of the Environmental Protection Agency, the Keeling curve caught the attention of a lot of people—including a Harvard undergraduate named Albert Gore, a student in Roger Revelle's population dynamics course—and played a key role in launching research programs into the effect of rising CO_2 on climate. (Not long after this, in the mid-1970s, I accepted an invitation from Revelle to join him and other Indian scientists aboard a research ship in the Arabian Sea, where we collected various ocean measurements. Global warming did not come up.)

By the time I left my village for good, the US government had become concerned about the societal and economic implications of global warming. To help guide future policy, the National Academy of Sciences (NAS) turned to the giants in the field of climate science—my mentors.

⁜

The first time I declined to study climate change, it was 1973 and I had just arrived at Princeton as a visiting student from MIT. I listened to Suki Manabe tell me about his latest GFDL-model simulation, the one in which he doubled the concentration of CO_2 in the atmosphere, becoming the first scientist to successfully model global warming. As staggered as I was by his results—more than 2 degrees Celsius increase in temperature—I had scientific eyes only for monsoon forecasting. I chose to use Manabe's model to examine the effects of the Arabian Sea's surface temperature on Indian monsoon rainfall.

Many years later, Manabe received some blowback for not making a bigger deal about those results, but my friend insisted that he was a scientist, not an activist. Besides, Manabe's model might have depicted a frightening future, but it didn't offer insight on whether humans might be able to adapt or what kinds of policies should be enacted to reduce emissions. The other thing it didn't provide was unequivocal confidence. Uncertainty is an unfortunate but unavoidable aspect of climate modeling; if there's one thing scientists are trained not to do, it's downplay or dismiss uncertainty.

But that doesn't mean Manabe stopped working on the problem. Rather, he ran more experiments and published more papers about the existential hazard of climate change. In 1977, he took part in the clumsily named Carbon Dioxide Effects Research and Assessment Program, meeting with seventy-five other scientists at a Miami Beach hotel to discuss the likelihood and threat of global warming. Their report pressed for further study and declared the importance of developing more sophisticated models, but on the whole, it addressed the risk in a decidedly dispassionate manner:

> At this time, scientists are not absolutely certain that significant climate changes will occur if the burning of fossil fuel continues or that the predicted climate change would, on the whole, be adverse. On the other hand, all climate changes will impose stress and we cannot ignore the possibility of long-lasting undesirable climate changes. It is imperative that society be able

to anticipate the consequences of fossil-fuel consumption and CO_2 release.[4]

In 1979, the Carter administration seemed to heed that advice when it asked the National Academy of Sciences to investigate and assess the scientific basis for human-caused climate change, appointing Jule Charney to lead the charge.

In July of that year, Charney invited eight of the country's leading climate experts to Woods Hole, Massachusetts, where they hunkered down to review all the available science on the effects of greenhouse gases. During the meetings, Charney called James Hansen, a NASA physicist who, with Manabe, had run model simulations for the project, and put him on speakerphone to discuss his results.

Hansen worked at the Goddard Institute for Space Studies in Manhattan, an office I had visited frequently with Milt Halem. A quiet and kind man with deep lines that bracketed his mouth like parentheses, Hansen studied the way carbon dioxide had changed the climate of Venus. Nothing about him—his demeanor or his research—foretold the outspoken climate activist he would one day become.

From his office in New York City, Hansen told the scientists on Cape Cod what his model had shown. If the amount of CO_2 in the atmosphere doubled—which, based on the Keeling curve, it was on track to do by the early twenty-first century—global temperatures would soar by 4 degrees Celsius, almost double the prediction Manabe had made a decade earlier.

Despite being pressured by the National Academy of Sciences to provide a precise number for global warming, Charney recognized the implications and insisted on using the two models, Manabe's and Hansen's, to name a range; he felt that indicating the science was certain down to a precise number would be misleading to the public. (Another committee member, Robert Dickinson, was also a member of another NAS committee being chaired by the great statistician John Tukey and he supported Charney's decision, saying that Tukey had emphatically argued that estimates must be given by a range, not a precise number.)

The final report predicted a warming in the range of 1.5 to 4.5 degrees, explaining that the range reflected "both uncertainties in physical understanding and inaccuracies arising from the need to reduce the mathematical problem to one that can be handled by even the fastest available electronic computers."[5] (This range of global warming suggested by the Charney committee more than forty years ago remains valid.)

Just like Manabe, Charney was committed to total transparency, but the work he did on what became known as the Charney Report convinced him that climate change was a real threat, and this time, he and the committee did not equivocate.

"It appears that the warming will eventually occur," the authors wrote, "and the associated regional climatic changes so important to the assessment of socioeconomic consequence may well be significant."

In a foreword to the report, the chair of the NAS's Climate Research Board was even more blunt: "If carbon dioxide continues to increase, the study group finds no reason to doubt that climate changes will result and no reason to believe that these changes will be negligible . . . A wait-and-see policy may mean waiting until it is too late."

The second time I declined to study global warming was when my adviser told me that global warming and climate change were going to be "a big deal in the future," as if he was urging me to think about the topic for later research. Again, dynamical seasonal prediction was my obsession; I didn't have the time or energy to direct my efforts elsewhere. But there was something else at play too, a gnawing sense that perhaps the observed global warming so far was not large enough to validate the fidelity of climate models and their future projections of global warming.

When I read the Charney Report, I couldn't help but feel uncomfortable with the committee's unapologetic reliance on the models, models that I knew firsthand to be, at times, inelegant, incomplete, and maddeningly unrefined. For instance, the two models used in the study produced two different results, one twice as big as the other. Why did they not insist on estimating the sensitivity of their results to changes in parameterizations? Why did they not explain why the models gave such wildly different answers? What if the models' global warming disappeared by some small change in one of the parameters? Who was to say a third model wouldn't have negated them entirely? We were only a decade out from the first successful one-day forecast and yet we were comfortable projecting decades into the future? There were so many sources of uncertainty I could not ignore.

I broached my concern with Charney himself. "Why didn't you include information in the report about how deficient and uncertain these models are?" I asked.

We were in his office at MIT; I had come to visit. He sat there for a moment looking at the Charles River and thinking. "I did some back-of-the-envelope calculations," he told me with his usual bravado, "and I knew it to be true based on my own understanding of the physics and dynamics of the atmosphere and oceans."

I decided to leave it at that. After all, I wasn't skeptical about the science but about whether we had the data and modeling to back it up. As the decade wore on, politicians—both Republican and Democrat—as well as some in the fossil-fuel industry accepted the report as settled fact, and more and more colleagues I respected, including Manabe, told me they trusted the models. My great respect for the opinions of Charney and Manabe nudged me away from my doubt and toward something more like ambivalence.

During the 1980s, as my fellow scientists and I plugged away at seasonal prediction at the University of Maryland, twenty minutes down the road, on Capitol Hill, politicians held debates and hearings about climate policy. Many of them were initiated by Al Gore, now a congressman and later a senator and a member of the committee on science and technology, and many featured James Hansen, who was becoming increasingly outspoken about the threat of man-made climate change.

Meanwhile, additional studies and investigations from esteemed agencies and NGOs corroborated what Charney had found in 1979, and finally, in 1983, the National Academy of Sciences' follow-up to the Charney Report was released. Meant to provide an exhaustive evaluation of the science thus far and suggest solutions for moving forward, the report, called "Changing Climate," failed to offer much in the way of new information. Worse, its lead author, William Nierenberg, a Scripps Institution physicist, downplayed its more damning findings when he spoke to the press and emphasized the approach he offered in the document's preface: "Our stance is conservative: we believe there is reason for caution, not panic."

Nierenberg's words had the impact of a wet towel on a campfire; for the next few years, no substantial action was taken toward establishing a climate policy, and the oil and gas industry quietly turned away from the conversation. It would take another atmospheric emergency—the hole in the ozone layer—for the conversation about greenhouse gases to resume. But by the time it did, attitudes among fossil-fuel executives had shifted.

The fossil-fuel industry had suppressed the warnings by their own scientists and surreptitiously launched a systematic campaign to create doubts about climate science in public and among policymakers. The industry recruited some scientists with questionable motivations who were willing to make public statements about the uncertainty of climate science. The conviction that global warming was settled science had evaporated among many Republicans in Congress. Politics and profits, it seemed, had entered and were sometimes driving the conversation. It was around this time that scientists began to report having their congressional testimony censored or outright changed by the White House.

But Hansen refused to be muzzled. In 1988, on one of the hottest days in a record hot year, Hansen once again appeared on Capitol Hill. Wearing a blue jacket and a red tie, the once-demure scientist boldly stated, "Global warming is now sufficiently large that we can ascribe with a high degree of confidence a cause-and-effect relationship to the greenhouse effect." Today, there is nothing radical about this statement, but at the time it was a bombshell.

I was in Australia when Hansen gave his testimony; I saw bits and pieces of it on the news. As I watched, I couldn't help but shake my head. *This guy is going to make our lives very difficult*, I thought. The next day, *The New York Times* landed on doorsteps all over the country with the headline "Global Warming Has Begun" splashed across the front page.

Mostly, I was concerned that the available observations and model results were insufficient to make such a significant statement. What worried me even more was that when I talked to experts on observations, they referred to confirmation from models; when I talked to my modeling friends, they referred to confirmation from observations. What if the global temperature, for some reason we couldn't anticipate, dropped? The fossil-fuel industry would surely go into overdrive to say that the science wasn't just uncertain, it was wrong. How would that reflect on scientists? I thought it best we shouldn't be too vocal until we knew for sure. In hindsight, I was wrong, and Hansen was right.

What followed was the acceleration of the rampant politicization in the United States that currently characterizes our climate crisis. On one side, existential panic. On the other, deflection and denial. Oil and gas companies went to war, endeavoring to "emphasize the uncertainty in scientific conclusions," as one Exxon paper urged. The American Petroleum Institute established a massive lobbying effort, recruiting scientists to express skepticism about global warming in the press. John Sununu, a fellow MIT alum and George H. W. Bush's chief of staff, had a toy climate model installed on his computer so that he could study all the ways that climate scientists—fueled by their bias against growth and development—had botched their projections.

Once, on a flight from Denver to Washington, I happened to be sitting next to former energy secretary James Schlesinger, who had recently published an op-ed in *The New York Times* claiming that the real reason European countries appeared to trust the science and supported a reduction in fossil-fuel consumption was their secret agenda to weaken the US economy. This was perhaps the most amazingly ridiculous statement from a senior-level official that I had ever heard (a record that would not stand for long). During the flight I asked him about it, certain he would demur and admit it had been a rhetorical tactic, but to my surprise the former secretary doubled down on the absurd opinion. It was then that I realized just how polarized the country was becoming over the issue.

Finally, there was action at last. Hansen's testimony galvanized scientists who were ready to take a more public stand in the face of global warming and collaborate with policymakers on solutions. By the end of 1988, the United Nations established the Intergovernmental Panel on Climate Change, the IPCC, which would become the global voice on climate change and was charged with producing the equivalent of the Charney Report every six to seven years.

During the 1990s, I had been asked several times to serve on the IPCC assessment-report committee but I had declined each time. I was busy; my travel docket was full. To work on an IPCC report required trips all over the world (the joke among climate scientists was that IPCC *itself* was responsible for much of the carbon in the atmosphere) and a grueling schedule of reading and meetings.

But in 2004, still awash in grief after Chandran's death, I received an invitation to serve on the IPCC's physical science committee for the sixth assessment report.

This time I said yes.

THE NOT-SO-FAMOUS WOMEN OF CLIMATE SCIENCE

Earlier in this chapter, I told the story of Svante Arrhenius and Guy Callendar and how scientists discovered the relationship between carbon in the atmosphere and the temperature on Earth. But as I myself discovered just a few years ago, that account wasn't quite right.

On August 23, 1856, Joseph Henry, an expert in electromagnetism and head of the newly formed Smithsonian Institution, stood before the annual meeting of the American Association for the Advancement of Science, held that year in Albany, New York, and read a paper entitled "Circumstances Affecting the Heat of the Sun's Rays."

In it, Henry described a simple but elegant experiment in which two glass cylinders were equipped with thermometers, filled with two different kinds of gases—including dry air and moist air and hydrogen and carbon dioxide—and left out in the sun. The result? The cylinder with moist air became warmer than the cylinder with dry air, and the cylinder with CO_2 became warmer than the one filled with hydrogen. What's more, when it was removed from the sun, the cylinder containing CO_2 took longer to cool.

"An atmosphere of that gas would give to our earth a high temperature," Henry concluded. "And if, as some suppose, at one period of its history, the air had mixed with it a larger proportion than at present, an increased temperature . . . must have necessarily resulted."[6]

This statement is notable for two reasons. First, it was one of the earliest recorded predictions of climate change. Second, it was written by a woman. For reasons that are unclear but probably easy to guess, Eunice Newton Foote, an American scientist and women's rights advocate, did not give this presentation, despite having thought up the experiment, performed it, and written the paper.

Perhaps that's why Foote's name and this experiment—conducted in her Saratoga Springs home and motivated by her own curiosity and desire to be useful to science—has largely been lost to history. Despite the fact that her pioneering work took place several years before that of the man who has long been credited with discovering the greenhouse effect (Irish physicist John Tyndall), several decades before Swedish scientist Svante Arrhenius made his observations on carbon dioxide, and almost a century before Guy Callendar connected the increase of atmospheric CO_2 with global warming, most people—including myself—hadn't heard her name.

Shamefully, this is not an uncommon phenomenon in science. (Not to mention society at large. In my own house, my parents gave a much higher priority to my brother and me than to our two younger sisters, who were married off as young teenagers.) Who knows how many other women—scientists, amateur scientists, students, lab assistants, or the overlooked wives of famous scholars—contributed to the field of climate science without recognition. Klara Dan von Neumann, wife of John and a Hungarian computer programmer, for example, rarely gets credit for her work programming the ENIAC, the electronic computer that produced the world's first twenty-four-hour weather forecast in 1950.[7]

In my own country, few know the story of Anna Mani, a brilliant scientist who defied traditional Indian gender roles and did much to professionalize the nation's meteorological agencies. Born in 1918 into a matriarchal state on the southern tip of India, Mani enjoyed a privileged upbringing and access to a large local library, the books of which she had plowed through by the age of twelve.

It was right around this time that Mahatma Gandhi visited her hometown, where he advocated for India's independence and called for a boycott of foreign goods. This idea would come to define Mani's tenure at the meteorological department, where she was in charge of the agency's instrumentation—as well as 121 men. In the early 1950s, Mani worked to end the import of meteorological instruments, simple tools to measure temperature and pressure in the upper atmosphere using balloons, and helped the country begin to manufacture its own.

Understanding that "wrong measurements are worse than none," Mani insisted that each instrument had to be carefully and accurately designed, constructed, calibrated, placed, and read.[8] The forward-thinking scientist would go on to design her own instruments to measure atmospheric ozone and solar radiation. It turned out that much of the instrumentation produced significant errors, but only later—with the help of advanced models—would scientists come to realize that.

Mani was deputy director general for instruments when I worked in Pune, and on her visits there, she used to occupy the guest room next to my office where I spent many late nights. We spoke only a few times, but I remember her kindness quite clearly. Much later I was surprised to learn that she had a reputation as a demanding boss and many of the men she supervised were scared of her!

In interviews later in her life, Mani revealed that she had faced a predictable amount of discrimination in her career, had been impugned for small mistakes and kept on the periphery of scientific debates. But she also rejected the notion that being a woman was what made her notable. "What is this hoopla about women and science?" she told one reporter. "My being a woman had absolutely no bearing on what I chose to do with my life."[9]

While gender equality in the field has yet to be achieved (male scientists made up more than two-thirds of the latest IPCC assessment report), today, thankfully, many of the seats of my classes in climate science are filled with women, a trend my colleagues in other STEM fields are also witnessing. And Eunice Newton Foote—her reputation rescued from the dustbin of history by female historians and championed by female climate scientists—now features prominently in my Intro to Climate Change PowerPoint presentation, a couple of slides ahead of Tyndall, Arrhenius, and Callendar, where she belongs.

Sixteen

My conversion from ambivalence to acceptance of global warming happened quickly. In an air-conditioned auditorium, surrounded by almost five hundred scientists representing nearly three-quarters of the countries on earth, I sat in a dumbfounded silence as expert after expert after expert after expert took the podium and delivered the sobering news from their corners of the scientific world like so many eulogies for life as we once knew it. It was merely the very first plenary session, and we would spend years diligently analyzing and vetting everything that had been published regarding climate change since the last IPCC assessment report. But before we even got started, an ominous feeling came over me.

Not many people realize that IPCC is a truly international effort, commissioned by the governments of the world *for* the governments of the world "to provide policymakers with regular scientific assessments on climate change, its implications and potential future risks, as well as to put forward adaptation and mitigation options," according to the group's mission statement. As such, each participating country—130 of them the year I was nominated by the Bush administration to join—sends its own scientists, all of whom then collaborate for many years to produce a massive report that serves as the scientific consensus on the state of climate and its future change—until the next report.

In short, every IPCC assessment report—which is produced every six to seven years—is like the Charney Report on steroids. That's because in 1979, when Charney and his pals met in Woods Hole to review the science around global warming, there were only a handful of papers to flip through. Between the IPCC's third and fourth assessment report alone, there were six thousand peer-reviewed papers published about climate change. We had our work cut out for us.

After that grim plenary session, we split into teams. There are three working groups for each IPCC assessment report: Physical Science; Impacts, Adaptation, and Vulnerability; and Mitigation.[1] I was in the first working group, and we were required to review the papers published in the peer-reviewed journals since the last report and examine the scientific underpinnings of the past, present, and future of climate change (the *why*). The second group would explore what effect a changing climate was having on society and ecosystems around the planet (the *how*). The third group would propose strategies to limit or reduce the amount of greenhouse gas emissions (the *what now?*). After rigorous review and debate, each group would prepare its own report, about the size of a phone book, representing the depth and breadth of human knowledge in that area.

The physical science report generally has about ten to fifteen chapters, and each chapter has about a dozen scientists assigned to it, all called lead authors. I was a lead author on the modeling chapter, "Climate Models and Their Evaluation."[2] During our first meeting, we divvied up the topics we would work on. In a fit of generosity, I told my fellow lead authors that they should choose the topics they really wanted, and I would take the topic that was left over. Perhaps I had assumed I would be assigned the section on monsoons, and I was a little disappointed when a fellow Indian scientist eagerly volunteered for it. That was how I ended up writing the section on tropical cyclones and hurricanes, not a frequent topic of my research. In fact, the last time I had consumed a trove of literature about tropical cyclones was during my first few days in Pune, when I'd stayed up many nights in a row cramming for my presentation.

It was an enlightening crash course, and ultimately, the most recent scholarship revealed that the models could not reliably predict whether the number of cyclones, hurricanes, and typhoons would increase or decrease each season. What seemed more reliable was that the severity of tropical storms would increase as the world, and especially the oceans, got hotter.

But just how hot would the world get? Even though each IPCC assessment report produces thousands upon thousands of pages, arguably the most important and most talked-about finding is how much Earth's surface temperature would increase across a range of scenarios—from what would happen if humanity stopped emitting greenhouse gases right this second to what would happen if humanity churned out even more. Coming to an agreement on this figure is one of the jobs of the Physical Science Working Group. To do so, we would be assessing simulations by twenty-three climate models from around the world.

In the name of so-called model democracy, IPCC treats all of these models equally, no matter how accurate or inaccurate they've proven to be in the past. The reasons for doing so are both political and practical. Because the project cannot be successful without the wholesale collaboration of every participating country, the scientific expertise out of, say, Russia, should be treated the same as that which emerges from the United States. There's also an accepted notion underpinning model democracy: just because a model performed well in one aspect is no guarantee it will perform as well in another. Different models have different strengths, so they should all be averaged together.

While I understood the political reasons behind model democracy and the reluctance of IPCC to rank the models for their fidelity and reliability, I never could quite get behind it. And when it came to the existential threat we had all gathered to consider, I felt strongly that model *meritocracy* would be much more appropriate. I didn't want to hurt anyone's feelings, but I certainly wanted to give humanity the best chance at survival.

So in 2006, as my IPCC duties were ongoing, COLA colleagues Tim DelSole, Mike Fennessy, Jim Kinter, Dan Paolino, and I designed a study that would satisfy my desire to rank those twenty-three models and find out if there was a relationship between how good each model was and how much global warming it predicted. In essence, we would examine the performance of each model in simulating the past one hundred years for which we have observations and see how accurately it recreated the past conditions. Then we would see how much future warming that model predicted. We had absolutely no basis to know what the result of this study would be.

When we analyzed the data, we saw that the quote-unquote best models consistently predicted greater degrees of warming, to the tune of 4–5 degrees Celsius, a catastrophic rise in temperature. It was a very simple calculation with a very frightening result, and one we thought the world would like to know about.

As it turns out, we were wrong on that last point. When we sent our paper—which warned that "projected global warming due to increasing CO_2 is likely to be closer to the highest projected estimates among the current generation of climate models"—to *Science*, one of the reviewers argued against including it in the journal, saying it would cause panic. This reviewer, a scientist of great reputation, called me a few days after its rejection to tell me it wasn't just society he was worried about but my own well-being. Climate deniers would make my life miserable for authoring such a paper, he warned.

The paper was readily published in the journal *Geophysical Research Letters* in 2006. Since our IPCC team had the responsibility to assess all published papers, I was sure we would highlight this rather significant result in our report. Instead, a fellow lead author, a scientist from the US Department of Energy, objected. He argued that we might be accused of privileging IPCC members. Suddenly, it felt like I had switched places with the scientists I had once thought too alarmist.

I felt like I couldn't get anyone to listen to me. (Looking back, the conclusions of the papers have stood the test of time.)

I didn't need to worry for long, because regardless of which model was the best, by 2006, all of them agreed: global warming was real, it was caused by human activities, and the planet was in trouble.

All IPCC models follow a consistent experimental design. Each model is run twice. The first is called the natural run, and this means that the model is run with levels of CO_2 naturally present in the atmosphere during the past one hundred fifty years. The models, like nature, show the ups and downs of global mean temperature that have nothing to do with humans—phenomena like El Niños and volcano eruptions. Each model gives a different range thanks to sources of uncertainty (such as clouds) that each treats a little differently.

For the second run, the models use the CO_2 that is *projected* to be released by humans into the atmosphere. Since it is not possible to predict future human activities and future breakthroughs in technology, IPCC analyzes several possible scenarios, from aggressive climate action to business as usual. Again, thanks to the uncertainty in the models, they produce a range of global warming for each scenario.

During the creation of our reports, for the very first time since the IPCC had been meeting, the differences between the two model runs—with and without greenhouse gases—had diverged enough that IPCC could confidently say that human activities were responsible for climate change.

Back in 1988, I had watched Jim Hansen declare that human-caused climate change was here, and I had serious concerns about the public's perception of the integrity of climate science. Now I was convinced that Jim Hansen was right. Finally, the models were sophisticated enough. Finally, I had the proof I had waited for. Finally, a signal had emerged from the noise.

All of us agreed on the science, but how to communicate these results posed some challenges. We were almost positive about our model simulations, but models *always* include some uncertainty—how to convey that? We were also almost positive that humans were the cause of global warming, sea-level rise, hotter oceans, and heavier rains, but of course it was impossible to know if there were factors we weren't aware of. When it was time to draft the actual report, tensions ran high. There were folks like me, people who wanted to state we were sure, and others who, for one reason or another, wanted to be more conservative. (While many not involved with the IPCC think of it as a liberal enterprise that exaggerates the extent of global warming and its impacts, in reality it is a conservative operation because all authors from various countries have to agree to the final conclusions.)

To address all the ambiguity inherent in the process, IPCC relies on confidence levels. An outcome that is 99 percent likely to occur, for instance, is *virtually certain*. Ninety-five percent? *Extremely likely*. There are ten of these terms, all the way to *exceptionally unlikely*, a phrase that describes an outcome with less than 1 percent probability.[3] Often, using these terms felt like a very technical way to describe a disturbing matter—like calling the wreck of the *Titanic* a *maritime mishap*—but it was the best we could do.

In 2007, the Physical Sciences Report, which comes out before the other two, was released to the public. Somewhere near the top, it plainly said:

> Warming of the climate system is unequivocal, as is now evident from observations of increases in global average air and ocean temperatures, widespread melting of snow and ice and rising global average sea level.[4]

And:

> Most of the observed increase in global average temperatures since the mid-twentieth century is very likely due to the observed increase in anthropogenic GHG concentrations.

And:

> Discernible human influences extend beyond average temperature to other aspects of climate. Human influences have: very likely contributed to sea level rise during the latter half of the twentieth century, likely contributed to changes in wind patterns, affecting extratropical storm tracks and temperature patterns, likely increased temperatures of extreme hot nights, cold nights and cold days, more likely than not increased risk of heat waves, area affected by drought since the 1970s and frequency of heavy precipitation events.
>
> Anthropogenic warming could lead to some impacts that are abrupt or irreversible, depending upon the rate and magnitude of the climate change.

While the IPCC fourth assessment report was the first to confirm that human activities were responsible for climate change, an assertion that made the front pages from Delhi to Washington, DC, and everywhere in between, frustratingly little action came from it. Politics had calcified the argument into a virtual stalemate, and meaningful policy would take years to come.

But for me personally, being a part of the IPCC report was transformational. I started as something of a skeptic and came out the other side as a strong believer.

However, nearly twenty years later, many of us in the climate science community no longer feel like so much human and computing resources to prepare IPCC assessment reports are necessary. Compiling one takes an extraordinary amount of time and effort, not to mention shameful amounts of travel (during my tenure on the panel, I flew to Norway, Italy, and New Zealand for IPCC meetings). Almost all climate-modeling centers prioritize and often stretch their resources to develop and test their models so that they can submit their results to the next IPCC panel, rather than improving their models. The IPCC calculations consume so much of the scientific and computational capacity that other important research topics—such as, yes, seasonal prediction—languish. More important, the IPCC reports continue to say the same thing over and over again. There is no good news that is going to pop out of the data—unless we actually do something about it.

A few months after our report came out, I was enjoying my tea on a quiet morning. I hadn't yet turned on my phone or my computer. I was content to just sit and think for a while. From the kitchen, I heard an "Oh!" from my wife.

"Did you know that IPCC got the Nobel Peace Prize?" she said, holding up her cell phone, a news alert stretched across its home screen.

"No," I responded, slowly realizing what this meant. "But . . . I'm a part of IPCC."

As my electronics fired up, what sounded like a hundred pings emanated from them: letters of excitement and congratulations from friends and colleagues. The IPCC would be sharing that year's Nobel Peace Prize with Al Gore:

> For their efforts to build up and disseminate greater knowledge about manmade climate change and to lay the foundations for the measures that are needed to counteract such change. Thousands of scientists and officials from over one hundred countries have collaborated to achieve greater certainty as to the scale of the warming.

Because the award was given to the IPCC as an institution, one that consisted of hundreds of scientists, it is not appropriate to call us Nobel laureates—a tiny concession. (That did not stop many IPCC members, especially their national news channels, to declare that the country had new Nobel laureates.) However, we all received colorful certificates from the IPCC "for contributing to the award of the Nobel Peace Prize for 2007."

How surreal it was to have my name associated with the biggest, most prestigious award in society for such a solemn discovery. After all, how could we celebrate if, as we wrote in the report, "unmitigated climate change would, in the long term, be likely to exceed the capacity of natural, managed, and human systems to adapt"?

SLAVES AND MASTERS

One of the reasons why our IPCC models couldn't quite agree on just how hot the planet would get was due to our lingering uncertainty about clouds. For many years, understanding how those puffy white things influence our climate has seemed as impossible as, to quote *The Sound of Music*, catching a cloud and pinning it down. To understand why they're so confounding, it helps to have a basic understanding of how clouds are born and the many multitudes they contain.

Think about the last time you looked out the window of an airplane as it made its ascent. Did you notice how the approaching clouds all seemed to rest on the same invisible shelf? Did you hold your breath as the plane pushed its nose through that shelf as your clear blue view turned fuzzy and gray? You had just crossed the line that meteorologists call the lifting condensation level—the point in the atmosphere where the temperature is cool enough to condense moisture into a cloud.

Where did this moisture come from? Look around. The ground, the trees, oceans, lakes, ponds, your pot of pasta water—all of it contributes to the moisture in the air. There is abundant moisture in the lower levels of the atmosphere; however, clouds form only over those regions where the moist air can be elevated to the lifting condensation level. To understand how large-scale cloud systems are formed, we must go beyond the thermodynamics of moist air (that it condenses at cold temperatures) and consider the dynamics of the atmosphere itself.

To make a large cloud system like the ones we see in formidable storms, a block of moist air—in climate science, we call it a parcel—needs to be lifted by large-scale vertical velocity created by large-scale dynamics of the atmosphere, such as the convergence of air toward the eye of a hurricane. As the parcel goes up, moisture in the air condenses, heating the parcel and sending it even higher since it is now lighter than the air around it. This process will go on and on until there is no moisture left in the parcel and it is no longer lighter than the ambient air. When exactly that happens depends on that day's atmospheric conditions, particularly the vertical profiles of temperature and moisture.

Additionally, the atmosphere is in constant flux. Inside a cloud, moisture is continually condensing and evaporating based on the temperature of the surrounding air. That's why a cloud that looks like a bubbling marshmallow one moment can look like a stegosaurus the next.

Because clouds are completely beholden to atmospheric conditions, sometimes they are referred to as slaves of dynamics. However, once they have formed, they have a profound effect on radiation and precipitation, making them more like masters. Globally, clouds in our current climate reflect 30 percent of the total incoming solar radiation. That is why it is said that the Earth's albedo is 0.3. Even today's most sophisticated models struggle to get the horizontal and vertical distributions of clouds and albedo right in order to make sense of the outsize impact clouds have on our planet's climate and on climate change.

As you surely know by now, climate is determined largely by a balance between the incoming solar energy and outgoing long-wave radiation. Clouds play a dual role in this process. They reflect solar radiation away and cool the planet, but they also trap the Earth's radiation and keep it from escaping into space. The reason clouds are such headaches for climate modelers is that the net effect of these two opposite effects is difficult to get right. Simply put, it depends on how high and thick the clouds are and *where* they are. It gets even more complicated when we consider the structure and microphysics of clouds: how much water is inside, how big the droplets are, what percentage have frozen into ice crystals, and so on. To include all of this information in a complex global climate model remains a challenge. Different modeling groups have come up with different formulations to parameterize, or treat, these processes, which is one of the reasons why they get different results for the current and future climate.

Clouds have also complicated our understanding of climate change. As the temperature and moisture of the planet continues to increase, so will the large-scale dynamics that create the vertical velocity needed for cloud systems. This means that as our immediate environment changes, as some plants die out or new insects move in, the clouds above our heads—how many there are and where they form—will change too.

Net global warming will be in part determined by this new distribution of clouds, which in turn will be determined by the new distribution of ocean conditions and the large-scale dynamics of the coupled ocean-land-atmosphere system. (For example, since the land gets warmer than the oceans, the air over land can hold more moisture. Add to this that when the temperature increases by 1 degree Celsius, the air can hold 7 percent more moisture before condensing, and you'll understand why today we're seeing more frequent and devastating floods.)

The good news is that uncertainty is slowly dissipating as our models improve. The bad news is that it looks like clouds may not be on our side. The latest IPCC Assessment Report concluded that "the net effect of changes in clouds in response to global warming is to amplify human-induced warming, that is, the net cloud feedback is positive." Some scientists, unwilling to admit defeat, are proposing to seed the clouds to make them brighter so that they can reflect more incoming solar radiation.

Humans won't be the only species affected by shifting cloud patterns and they are hardly the only creatures that notice them.

In India, peacocks begin "dancing" when they see the first monsoon clouds appear on the horizon. Butterflies feed more on cloudy days.[5] Geese sleep less on cloudy nights.[6] In the ocean, zooplankton have been observed making "mini-migrations" during times of cloud cover, relying on the shadows provided by clouds to feed closer to the surface in relative safety.[7] And in the high elevations of the Southern Hemisphere, cloud forests—verdant with trees, mosses, ferns, lichens, and epiphytes—have evolved to enjoy life above the lifting condensation level.

Seventeen

For many years, it seemed that the only group spurred to action by the 2007 IPCC Assessment Report was the fossil-fuel industry. While the US government again and again failed to pass any meaningful legislation to curb emissions, oil and gas companies were spending nearly one billion dollars a year to sow doubt among the American public about the scientific certainty of human-caused climate change.[1]

Of course, we now know that these efforts had been going on for years. As early as 1998, the American Petroleum Institute (API)—a lobbying group that represents ExxonMobil, Chevron, BP, Shell Oil, ConocoPhillips, and others—was on the offensive, launching a public relations campaign called the Global Science Communications Action Plan. Its goal? That "a majority of the American public recognizes that significant uncertainties exist in climate science, and therefore raises questions among those (e.g. Congress) who chart the U.S. future course on global climate change,"[2] according to a memo leaked to

The New York Times. In it, the lobbying group outlined an ambitious plan that included recruiting a network of sympathetic scientists to appear across media platforms, establishing a nonprofit foundation in Washington to serve as an "alternative to the IPCC," and developing curriculum materials for schools that would present a "balanced picture" of climate science—a picture, no doubt, that minimized the role of oil and gas in global warming. After the IPCC and Vice President Gore received the Nobel Peace Prize, the fossil-fuel industry redoubled their efforts to misinform the public and the policymakers and create doubt about the science and impacts of global warming on future generations.

A decade and a half later, API hadn't achieved everything it set out to do in its 1998 plan, but it had made major progress toward its stated goal. Between the year the IPCC won the Nobel Peace Prize and the spring of 2015, *fewer* people (and especially fewer policymakers) believed that global warming was caused by humans burning fossil fuels.[3] As a climate scientist who has long been devoted to using the work to make the world a better place, I found watching such a large segment of society reject urgently important science a unique agony. For our IPCC report and those that came after, hundreds of the world's smartest people had put aside their own lives and scholarship to produce a rigorous and instructive document charting the best path forward. And over and over again, it was ignored.

Whenever I traveled to Europe or India, my friends and colleagues would ask why Americans didn't believe in global warming. I could never give a clear answer. Perhaps scientists like me—waiting for proof that was late in coming—had given the fossil-fuel industry too big of a head start. Perhaps we would never catch up. So on the morning of May 31, 2015, as I sat at the kitchen table reading *The Washington Post*, I was struck by the title of an op-ed written by Senator Sheldon Whitehouse of Rhode Island: "Big Oil Seems to Be Acting Like Big Tobacco." The senator's letter pointed out the striking "parallels between what the tobacco industry did and what the fossil-fuel industry is doing now," including creating a network of so-called experts to push back on the validity of accepted science while cashing in on a catastrophically destructive product. Whitehouse argued that the Justice Department should do what it had done with tobacco companies in the 1990s and file a civil RICO lawsuit (which allows the government to go after groups involved in organized crime) to uncover the vast misinformation campaign orchestrated by the oil and gas industry. It was only after the RICO inquiry that the tobacco industry publicly accepted the scientific findings that nicotine was addictive, and smoking was injurious to health. Scientists working for the tobacco industries had come to this conclusion much earlier, but the leaders of the industry had suppressed the results.

As the late-morning sunlight moved across the newsprint, I sat back in my chair and thought about Senator Whitehouse's idea. It was, I believed, a very good one. Maybe I could lend my voice, not just as a climate scientist but as a private citizen, to encourage those in power to consider it. I decided I would engage in one of the oldest and most sacred traditions of American democracy: I would write a letter to my representatives. Perhaps I could convince some of my friends and colleagues to sign it too.

That afternoon, I opened up the blank draft of an email and began typing names as they came to me of close friends in my community, especially those who had a robust understanding of climate science but had never spoken out about climate change. Would they be willing to send a letter of support to the senator and others? I wanted to know. The replies were swift and almost unanimously positive (only two declined). As the online conversation developed, the idea took on a new shape. One of my GMU colleagues suggested that in addition to endorsing the senator's idea, we should send our letter to the attorney general, the president's science adviser, and the president himself. Not everyone contributed to the text of the letter, but the final text was approved by all twenty of the signatories.

On September 1, 2015, before I left for work, I asked Anne if she would mind dropping off the following letter in a mailbox on her way to the grocery store.

Dear President Obama, Attorney General Lynch, and OSTP Director Holdren,

As you know, an overwhelming majority of climate scientists are convinced about the potentially serious adverse effects of human-induced climate change on human health, agriculture, and biodiversity. We applaud your efforts to regulate emissions and the other steps you are taking. Nonetheless, as climate scientists we are exceedingly concerned that America's response to climate change—indeed, the world's response to climate change—is insufficient. The risks posed by climate change, including increasing extreme weather events, rising sea levels, and increasing ocean acidity—and potential strategies for addressing them—are detailed in the Third National Climate Assessment (2014), Climate Change Impacts in the United States. The stability of the Earth's climate over the past ten thousand years contributed to the growth of agriculture and therefore, a thriving human civilization. We are now at high risk of seriously destabilizing the Earth's climate and irreparably harming people around the world, especially the world's poorest people.

We appreciate that you are making aggressive and imaginative use of the limited tools available to you in the face of a recalcitrant Congress. One additional tool—recently proposed by Senator Sheldon Whitehouse—is a RICO (Racketeer Influenced and Corrupt Organizations Act) investigation of corporations and other organizations that have knowingly deceived the American people about the risks of climate change, as a means to forestall America's response to climate change. The actions of these organizations have been extensively documented in peer-reviewed academic research (Brulle, 2013) and in recent books including: Doubt is their Product (Michaels, 2008), Climate Cover-Up (Hoggan & Littlemore, 2009), Merchants of Doubt (Oreskes & Conway, 2010), The Climate War (Pooley, 2010), and in The Climate Deception Dossiers (Union of Concerned Scientists, 2015). ***We strongly endorse Senator Whitehouse's call for a RICO investigation.***

The methods of these organizations are quite similar to those used earlier by the tobacco industry. A RICO investigation (1999 to 2006) played an important role in stopping the tobacco industry from continuing to deceive the American people about the dangers of smoking. If corporations in the fossil fuel industry and their supporters are guilty of the misdeeds that have been documented in books and journal articles, it is imperative that these misdeeds be stopped as soon as possible so that America and the world can get on with the critically important business of finding effective ways to restabilize the Earth's climate, before even more lasting damage is done.

Sincerely,

J. Shukla, E. Maibach, P. Dirmeyer, B. Klinger, P. Schopf, D. Straus, E. Sarachik, M. Wallace, A. Robock, E. Kalnay, W. Lau, K. Trenberth, T. N. Krishnamurti, V. Misra, B. Kirtman, R. Dickinson, M. Biasutti, M. Cane, L. Goddard, A. Betts

Soon after I reached my office that day, I was told that one of Senator Whitehouse's aides had left a message for me (his staffers had been given a heads-up about the letter by someone in the climate science community). When I returned the call, the staffer told me that the senator appreciated our support but wanted us to delay sending it because there was a much larger group—two hundred instead of twenty—that wanted to sign it. I quickly hung up, called my wife, and asked her to hold the letter. But it was too late. She had already gone to the local post office and mailed it. We talked about the possibility of retrieving it but decided to let it go.

Truthfully, I didn't think much would come from this letter. Mostly, I hoped it would satisfy some of the nagging guilt I felt over not doing enough to fight for the future of the planet our children and grandchildren would inherit. Little did I know that my fight was just beginning.

✳

The first trickle of blogs and articles about the #RICO20, as we would come to be called, was posted on September 17, 2015.

The *Daily Caller*, a right-wing news and opinion website founded by the now-former Fox News host Tucker Carlson, characterized us as "riled up academics . . . asking President Barack Obama to prosecute people who disagree with them on the science behind man-made global warming."[4]

A Fox News headline proclaimed "A New Low in Science: Criminalizing Climate Change Skeptics," while Breitbart summarized our letter thus: "Climate Alarmists to Obama: Use RICO Laws to Jail Skeptics." Its author noted that our "hypocrisy and dishonesty . . . almost defies belief."

A blog called *twitchy*, in a creative use of quotation marks, said (falsely) the group "wanted to criminalize 'views counter to theirs' and send to 'jail' those engaged in 'scientific dissent.'"[5] Another blog, written by a former Cornell professor, called us "cowards, inferior intellects, crybabies, poor losers, promulgators of a failed science."[6]

It was somewhat surprising to see the hosts of so many different channels and the authors of so many different blogs all repeating the same words as if they all had received the same memo. One evening I happened to be watching Bill O'Reilly of Fox News when he instructed his staff to publish the names of these twenty "criminals" who'd signed the RICO letter.

As the stories were published, people began bombarding my phone and email with vitriolic and abusive missives:

> *Take your ass back from wherever you came, lying piece of crap.*
>
> *Let me ask you something, "Professor." If the issue of climate "science" is now settled (according to you and the 19 other crackpots signatories to that letter to the Attorney General), why is it necessary to fund its research with any more grant money?*

Typical liberal sucking at the taxpayer teat.

Brown Shirts like you remind me of why my Jewish father fled Nazi Germany.

Your Nazilike mentality to stifle healthy debate on a far from settled controversy is preposterous. You need therapy and counseling.

What a glorious day it is to see the idiocy behind your RICO smear campaign coming to light!

I don't know what country you crawled out of, but here in the US, we have free speech.

Have a heart attack.

Jail ki hava khana [That's Hindi for "you will go to jail."]

I had never heard of, let alone read, most of the websites that were publishing this inaccurate and disparaging coverage and I wasn't about to start. I knew we had done nothing wrong by sending the letter, so I decided to keep my head down and stay out of the fray. But the butterflies of chaos had other ideas, and a tiny act was about to cause a tornado of trouble.

One of the signatories of the letter called my office to suggest we put the text on the IGES website so that there was a link that could be emailed and tweeted. I agreed this was a good idea; I wanted the letter to be shared as far and wide as possible.

At the time, IGES was essentially a defunct institution, and the process for formally dissolving the nonprofit had been underway for a year or more, the final step and a requirement in the transition of COLA from a private nonprofit to GMU. But since the website still existed, I thought it would be a fine place to post the letter. So I asked one of COLA's staff—who just happened to be walking down the corridor when I received the call—to put it online. I did not consult the head of IT or the director of COLA about my decision. I was so naïve.

It didn't take long to recognize my mistake.

Within hours, the tenor of the articles and blogs had shifted. Because I had posted the letter—a document I didn't find all that political but many did—on the website of IGES, a previously government-funded organization that I had founded, suddenly *I* was the controversy. Now the headlines were about me:

Climate Scientist Asking Obama to Prosecute Skeptics Got Millions from US Taxpayers

The Lead Climate Scientist Behind the Obama/RICO Letter Has Some Serious Questions to Answer

Getting Rich Off Climate Extremism

Leader of the RICO 20 Is a Prime Example of a Climate Profiteer

Climate Scientist Asking Obama to Prosecute Skeptics Made Millions Off "Double-Dipping"

Global Warming Double Dipper Enriches Family with Tax Dollars

Double-dipping. It was a phrase that would haunt my dreams for the next year. The allegation of these climate deniers was that, because I drew a portion of my salary from both George Mason University and IGES, I had committed an ethical and legal breach, that I was a climate-change grifter, using public funds to fuel unfounded fear over global warming to pad my own wallet. Worse, they said my wife was in on the grift, drawing a salary from our sham of a nonprofit.

In reality, my arrangement with GMU was standard practice among academics with significant research projects *and* it was carefully documented with contracts, memoranda of understanding, and meticulously kept time sheets. In reality, I had been something of a climate-change skeptic myself, for years hesitant to join the IPCC.

In reality, our letter never so much as hinted at imprisoning dissenting scientists. In reality, I had resigned my tenured professorship at Maryland and taken the risk of having no job if my research grant proposals to NSF, NOAA, and NASA were rejected. In reality, Anne was a founding member of IGES who had agreed to mortgage our home and start the institute in the garage, and she and I regularly gave away nearly one-third of our annual incomes, drove modest cars, had lived in the same home for decades, and sacrificed years of time together to bring life-changing science to the people who needed it most. In reality, I had sent the letter as a private citizen, and it had been mailed from a post office near my house; the return address on the envelope was my home. Suddenly, reality didn't matter. In the eyes of the misinformation machine, now I was the "Climate Alarmist Caught in 'Largest Science Scandal in U.S. History,'"[7] and the "Third Most Dangerous Person" of 2015, according to Fox News.

It was an irony not lost on me that I had coordinated the signatories of the letter in an effort to expose that misinformation machine at the center of climate-change denialism, and now I had the uncanny experience of standing in the center of its violent bull's-eye, getting a full and terrifying picture of just how vast and well financed the machine was.

On October 1, one month after we mailed the letter, I received a letter myself. It was from a congressman named Lamar Smith, chairman of the US House of Representatives Committee on Science, Space, and Technology.[8] Until that moment, I had no idea who Lamar Smith was. Crowded with footnotes naming those fringe blogs and right-wing websites, Smith's letter raised the specter of an investigation into why IGES, a tax-exempt nonprofit, was "participating in partisan political activity." The letter instructed me to preserve and be prepared to hand over years' worth of electronic and physical documents. (I was flabbergasted later to see that the inflammatory websites and bloggers seemed to know before I did about Mr. Smith's letter.)

That same day, the offices of NASA, NOAA, and the NSF also received letters from Congressman Smith asking for their records having to do with research grants to IGES.

It seemed like a good time to look for an attorney. When I met with one for my first consultation, I asked him if he thought I needed representation right then or if I should wait for an actual investigation to be announced. I'll never forget the grave look on his face as he advised me: "Dr. Shukla, it's best to have a lawyer *before* the FBI knocks on your door." It was not a comforting thing to hear.

Luckily, a GMU colleague suggested another attorney. His name was Edward Newberry, and he was a partner at Squire Patton Boggs, one of the top law firms in the world. I was impressed by his knowledge and understanding of legal matters like mine but even more impressed when he told me that he was reading *Superforecasting*, a book about uncertainty and probability in weather forecasting, which he said he found useful in discussing possible outcomes with other attorneys in his firm. After a short paid engagement, Ed Newberry and his colleague Clark Kent Ervin ended up helping me pro bono; it was my sheer good luck to work with such great attorneys.

A few months later, with Ed by my side, I took my seat in a conference room in the Rayburn House Office Building in downtown Washington, DC. With a stack of requested documents in front of me, I sat at the conference table facing about eight staffers of the Committee on Science, Space, and Technology. As they began their questioning, mostly about my role in the creation and administration of IGES, I felt calm and resolved. I challenged some of their assumptions and answered all their questions with facts and figures. The better I handled this firing squad, I thought, the easier it would be for the scientists who would no doubt be sitting in this chair in years to come. At the end of nearly two hours, they started looking at each other as if they had run out of questions. After the meeting, Ed was sanguine, assuring me that my testimony had been all the committee needed to clear up any genuine misunderstanding. I was certain that was true.

Unfortunately, both Ed and I were wrong. Behind the scenes,

Congressman Smith was still busy writing letters, and his office was sharing documents I had provided with editors at major newspapers. Meanwhile, various fossil-fuel-industry-funded organizations, like the Competitive Enterprise Institute (funded by Charles Koch Foundation and David Koch Foundation, an ExxonMobil cause[9]), and the Cause of Action Institute (a Koch brothers favorite[10]) were bombarding our universities with FOIA (Freedom of Information Act) requests, demanding the release of any emails that had to do with our letter. I was told that some Koch brothers' acolytes on the GMU board of visitors and a dean of the law school were urging GMU to investigate and fire the GMU professors who signed the letter.

That spring, just as the cherry blossoms began to open along the Potomac Tidal Basin, I learned that in response to letters from Mr. Smith, the National Science Foundation's Office of the Inspector General would conduct its own investigation into IGES, its finances, and me.

In hindsight, I'm not sure what's more unbelievable: the sheer quantity of documents that the NSF inspector general requested or that I was able to provide them with everything they asked for. A lifelong pack rat, I save every scrap of paper having to do with work, and it had finally paid off. When the NSF inspector general asked for biweekly time sheets for all employees of IGES and COLA for the past eight years, Anne delivered them several boxes of original time sheets signed in ink the very next day.

But just because the documents were accounted for didn't mean that there wasn't a small mistake contained within them. I had been filling out time sheets for more than twenty years, carefully documenting the time I spent on GMU work and the time I spent on IGES work. What if on one of those time sheets, I had inadvertently miscoded an hour or a day? I had no doubt, especially given the political pressure from Lamar Smith, that if the OIG (Office of Inspector General) found any discrepancies, they would call it evidence of double-dipping, and that would mean the end of my career. Thanks to meticulous record-keeping by Anne and our strict adherence to all federal regulations, after reviewing time sheets and paid invoices for the previous ten years, the NSF inspector general did not find a single discrepancy.

As the NSF requested more documents and sorted through reams of paperwork (wasting staff time and money that surely could have been used in a more productive manner), there was little for me to do but wait. I spent a lot of that time buzzing with anxiety. I lost weight and needed sleeping pills to fall asleep; my doctor was worried. Professionally, my life went on as it always had. I was invited to join committees, give lectures, attend conferences, and make presentations. In the middle of all that, I was informed that the American Meteorological Society had selected me to be an honorary member, the highest honor given by the society. But the investigation hovered on the periphery of my thoughts. I was bothered by the attack on my integrity, but more than anything, it was hampering my research, and I felt distraught about how any meaningful climate action would get done if this was the reaction to something as innocuous as a letter. Despite multiple threats, I refused to withdraw the letter.

I was also angry.

Over the course of the so-called scandal, I learned that during his career, Congressman Smith had received nearly eight hundred thousand dollars from oil and gas companies, making the fossil-fuel industry his biggest donor.[11] I can only guess this was why Smith was so aggressive in his capacity as chair of the House Committee on Science, Space, and Technology, and not just in the case of me and the #RICO20. The Texas congressman issued more subpoenas in his first three years than the committee had for its entire fifty-four-year history.[12] Between the politicians, websites, and so-called think tanks it funded, the fossil-fuel industry had amassed a formidable army.

It was an army, it turns out, that had already invaded what I had for so long assumed to be safe, neutral territory—science. Perhaps the most painful part of my tangle with the fossil-fuel industry was learning how far its influence reached—fellow scientists I once considered friends and leaders at institutions I had faithfully served for years not only failed to defend me but actively participated in trying to bring me down. Since leaving my village, I had always found refuge in the universal constants and unchanging laws of physics. I don't think it was a coincidence that I spent my career looking for predictability in the midst of chaos; I had spent my life looking for it too. Now, during many sleepless nights, I grappled with the loss of the stability my career had long provided, my life raft in a cruel and chaotic sea.

Sometimes, though, as the blackness of midnight gave way to the purple of morning, I found comfort in revisiting some of the most meaningful moments of that career. The hours spent in a freezing cold room setting up India's very first supercomputer; standing at a lectern in Geneva delivering the world's most accurate El Niño forecast; sailing low over the yellow grass of the Kansas prairie and the choppy waters of the Bay of Bengal.

It took me a while to realize it, but at some point during the eighteen-month investigation, those memories began revealing something else: It wasn't actually the science that had been my life raft. When I was a young man looking out of the window of that hurricane hunter, it wasn't making a more perfect model that drove me; it was making a more perfect world. Hope, optimism, and an obligation to help others; it was *that* commitment—not the science itself—that had been my terra firma all along.

It dawned on me that the real threat that Congressman Smith and the climate-denial machine posed wasn't that they would destroy my reputation or even invalidate my science. It was that they might snuff out that hope, that optimism, that commitment—in me and in others. For the future of my children and grandchildren, not letting that happen was a battle I was willing to fight.

It was this thought that sustained me through the early days of November 2017, when I received an email from the NSF. The *"conclusion is that there is no evidence to substantiate the allegations. This investigation is closed and no further action will be taken,"* it read. I was told it was one of the shortest reports ever issued by the organization's inspector general.

(Three days before I'd received the letter from NSF OIG, Congressman Smith announced he would not seek reelection, though there was no shortage of climate deniers on Capitol Hill ready to take his place.) When I forwarded the email to the family, friends, and colleagues who had supported me during the nearly two years of chaos, I told them it had been a small price to pay to defend the integrity of climate science.

The day I sent the email, I also thought for the first time in a long while about a phrase I had only ever heard in the village: *Sanch ko anch nahin*. It's a saying that describes the way to tell wood from gold—put them both in a blazing fire and see how they react. Literally translated, it means "Flames cannot affect the truth."

President Obama did eventually respond to our letter in April 2016, writing:

April 26, 2016
Professor Jagadish Shukla Rockville, Maryland

Dear Professor Shukla:

Thank you for writing. The threat of a changing climate holds the power to define the future of our planet more dramatically than any other. That is why, driven by the resolve and spirit of common purpose that have always guided American leadership, my Administration has taken bold action to ensure that we preserve our planet for generations to come.

The scientific consensus on climate change is clear. An overwhelming body of evidence supported by 99 percent of scientists shows us that the global climate has been rapidly changing in many ways. These changes are being driven primarily by human emissions of greenhouse gases, and the consequences associated with these trends will become increasingly threatening if we do not significantly reverse the pace of emissions. Recently, we have seen some of the impacts of climate change in the stronger storms, deeper droughts, and longer wildfire seasons that have caused so much damage in communities across America.

At home, our country has made unparalleled progress in confronting climate change by moving toward a low-carbon economy. Since I took office, our nation has multiplied wind power threefold, expanded solar power more than thirtyfold, and worked to increase access to affordable clean energy sources. We have established the first ever carbon pollution standards for power plants-the largest source of carbon pollution in our country—and my Administration has consistently worked with cities, states, and businesses to continue to boost our energy efficiency. Additionally, I worked with Congress to secure a long-term extension on renewable energy tax credits, and the budget proposal I sent to Congress this year would double the funding for clean energy research and development by 2020—including investments to help the private sector create more jobs and lower the cost of clean energy.

Abroad, the United States has led the international community's efforts on climate change. Last December, in Paris, nearly 200 nations came together to conclude the most ambitious climate change agreement in history. The Paris Agreement establishes a foundation for keeping the rise of global temperature below 2 degrees Celsius and sends a powerful signal to the private sector that the global economy is shifting to clean energy sources. While our initial goals are just a start, the Paris Agreement provides the architecture to craft more ambitious targets as we continue to make breakthroughs in clean energy technology.

Again, thank you for writing. Our generation has a unique responsibility to act on this issue with the urgency it demands while we still have time. As long as I'm president and beyond, I'll keep pushing to ensure our children inherit a secure and sustainable future. For information about how climate change is impacting the United States, visit nca2014.GlobalChange.gov. *To learn more about my action plan to address the threat of climate change, visit* www.WhiteHouse.gov/Climate-Change.

Sincerely,
Barack Obama

SILENCING SCIENTISTS WITH FOIA REQUESTS

I can tell you from experience: There is hardly anything more terrifying than an elected official alleging you have mishandled government funds—and there is nothing more *overwhelming* than being asked to turn over every piece of paper you have produced in the past decade. Guess what takes a back seat while you deal with those allegations and requests? Science.

I learned a lot from my tangle with the fossil-fuel industry and the vast network of climate deniers it keeps riled up and well funded. Most of all I learned that my case was hardly unique. In fact, intimidating scientists by questioning their use of public funds and requesting ungodly reams of correspondence is pretty much page one of the climate deniers' handbook. Often, it doesn't take much to be targeted in this way.

All Andrew Dessler, an atmospheric scientist at Texas A&M University, did was answer the phone when a journalist from *The New York Times* called. Dessler's name appeared twice in an article about climate deniers and their obsession with clouds. "If you listen to the credible climate skeptics, they've really pushed all their chips onto clouds," he said.[13]

The next day, Texas A&M received a FOIA request from Christopher Horner, who at the time was working for the Competitive Enterprise Institute, an ultraconservative think tank with connections to the Koch brothers.[14]

The Freedom of Information Act, or FOIA, was passed in 1966 to help the press report on the federal government and keep it honest. Through FOIA requests, journalists can get access to previously unreleased government documents. Today, though, less than 8 percent of FOIA requests come from news outlets.[15] More than a fifth come from individuals like Horner, who asked Texas A&M to turn over any of Dessler's emails that included the terms *Lindzen, Michael Mann, hockey stick, Climategate, denier,* and *tobacco*. (Some of these terms were completely unrelated to the *New York Times* article.) Horner wrote in the *Washington Examiner* that his organization was committed to investigating how "taxpayer-funded academics use their positions to advance a particular agenda."[16]

After Dessler appeared in an episode about the politics of climate science on the PBS program *Frontline*, the Competitive Enterprise Institute (CEI) filed yet another FOIA request.

Unsurprisingly, no legal action resulted from either FOIA request.

That same year, CEI targeted another climate scientist, Texas Tech's Katherine Hayhoe, filing three separate FOIA requests with the university seeking evidence that Hayhoe had misused public funding when she authored a chapter for a book about climate policy, a standard part of any academic's job. Nothing came of these burdensome requests—though in reality, legal action might not be the point.

According to a 2015 report produced by the Union of Concerned Scientists, "open records requests are increasingly being used to harass and intimidate scientists and other academic researchers, or to disrupt and delay their work . . . this may often be the main purpose of such requests."[17] I would also argue that these requests are made to sully the reputation of scientists, as merely the *idea* that someone might have misused public funds can raise suspicion or contempt in the minds of average Americans.

One of the most egregious uses of bad-faith FOIA requests is the well-known case of Michael Mann, an atmospheric scientist at Penn State. In the late nineties, he and a colleague worked to reconstruct the average temperature of the Northern Hemisphere for the past thousand years, producing a graph that depicted the sharp increase of the twentieth century, an image that became known as a hockey-stick graph.

In 2005, another Texas congressman, Joe Barton, commissioned an investigation into Mann's work, requesting funding sources, computer codes, and personal information. Five years later, Virginia's attorney general requested every email and document involving Mann from the University of Virginia, where the scientist worked from 1995 to

2005. When a judge ruled that the university did not need to comply with the request, ATI (now called Energy and Environmental Legal Institute) took up the fight, using a FOIA request to demand the same documents. The case made it all the way to the state's supreme court, which ultimately ruled in favor of Mann and academic freedom—a rare victory for climate science.

During the long trial, a diverse group of environmentalists came together to raise money to help defray Mann's legal fees, an effort that eventually morphed into the Climate Science Legal Defense Fund, an organization that provides legal expertise to climate scientists being harassed, censored, or intimidated. Lauren Kurtz and CSLDF were the first to support me. I remain very grateful for their support during my own ordeal. The defense fund continues to help defend scientists under attack from industry-supported groups.

While scientists are getting better at fighting back, the anti-science movement is making headway. In the past five years, due in part to rampant misinformation that characterized the pandemic, the number of American adults who have "a great deal of confidence" in the scientific community has fallen from 48 percent to 39 percent. Approximately 13 percent said they had "hardly any."[18]

It can be a difficult time to be a climate scientist—but it's a critical one.

Eighteen

In the fifty-five years I have been a part of the scientific community, researchers have discovered so many amazing facets about our planet's atmosphere, climate systems, and weather patterns—that the temperature of the ocean can have a dramatic impact on societies hundreds of miles away; that the seemingly unchanging land beneath our feet plays an integral role in how much rain falls on our heads. My colleagues and I have obsessed over the tiniest details of our climate models and supercomputers, dispatching countless balloons, ships, planes, and satellites to inform them. We've invested untold dollars and hours in the reanalyses of past weather and installed powerful forecasting infrastructure in countries that need them most. Now that I and the men and women I started my career with are reaching retirement, we are contributing time and money that will support the next generation of scientists in the form of scholarships, fellowships, and entire university departments. At this year's meeting of the American Meteorology Society, I was overjoyed to see the first presentation of the Jagadish Shukla Earth System Predictability Prize, an award given to a researcher who has made an outstanding contribution in—what else?—predictability, for the benefit of society.

And yet, despite all of the accomplishments of the past half century, I have always felt a nagging sense that there was much more work to be done, many more people to help, and always more letters to write.

While our letter to President Obama was perhaps the most public correspondence I have taken part in, I should confess now that I have written and continue to write a great many letters. I am my father's son; if I see a problem and have an idea of how it might be fixed, I do not hesitate to share my solution with those in power. Over the years this has earned me a reputation as being a bit of a bother, but it has also helped get a lot of important work done.

Writing a letter, in fact, is partly how I was able to convince my alma mater Banaras Hindu University, which awarded Ph.D.s to those working in research capacities at other agencies, that the Indian Institute of Tropical Meteorology, my employer, should be recognized as one of those agencies. I had a hunch that IITM had not been recognized because the institute was new—and because it was centered on what was considered at the time to be the dubious science of meteorology. So I wrote a letter arguing that scientists at IITM should be allowed to submit doctoral dissertations based on their research. A few years later, I submitted my dissertation and was awarded my first Ph.D.

At MIT in 1971, I attended a small gathering of Indian students held by the US ambassador to India, Triloki Nath Kaul. At the end of the meeting, the ambassador encouraged us to write to him if we had any ideas about how we might improve Indian society or relations between the two countries. I can't imagine many in the audience took the ambassador up on his invitation, but I sure did.

Since my days at IITM, I'd believed that the India Meteorological Department should do away with their policy of seniority-cum-fitness, which rewarded employees not for their accomplishments but for how long they'd worked at an organization and how much their bosses liked them. This system had led to some people woefully unqualified in meteorology heading up the entire department. For a time, in fact, IMD was run by a man whose qualifications included a career in seismology (the study of earthquakes)—and a lifetime at the department.

So I wrote Ambassador Kaul a letter. In it I argued that the head of the weather department should be appointed by merit. When nothing came from it, I wrote again. This time, the ambassador responded that he was forwarding my suggestion to the prime minister, Indira Gandhi herself. Soon, I was invited to New Delhi to speak before a commission that had been established to consider the issue; a few years later, the seniority-cum-fitness policy was scrapped, and merit-based appointments became the norm.

In the early 2000s, I dashed off a quick letter to India's prime minister about the country's outdated forecasting capabilities. I was a little shocked when I actually got a response—and an invitation to meet with him the next time he was in Washington.

That's how I ended up in a sitting room at the Blair House telling Prime Minister Singh, one of his senior ministers, and the deputy chairman of India's Planning Commission that "one of the world's greatest democracies has the most outdated weather service among developed economies," while the senior minister sat nearby, looking quite annoyed. "India has such potential," I added, mostly for this man's benefit. "And there are many scientists of Indian origin all over the world, experts of weather and climate, who would be willing to help."

The prime minister, a former economist who understood the importance of accurate monsoon forecasts for the Indian economy, peppered me with questions so long that one of his aides had to pull him from our meeting. "The director of the World Bank is waiting for you, sir," I was shocked to hear her say. When the prime minister asked for my card, I patted my suit pocket, horrified to realize I hadn't brought one. No matter. During my next visit to India, Singh invited me to his residence, setting in motion my heading up the International Advisory Panel for the Ministry of Earth Sciences, a brand-new ministry the prime minister had created soon after his return to India.

I highly recommend writing letters.

Of course, lots of letters do nothing but wind up in the trash can or gather dust on the corner of a secretary's desk. I have written countless missives arguing for new programs, new committees, new institutions. Some of them led to consequential outcomes; some were ignored or nothing useful came of them.

My multiple letters in the early 2000s to US federal agencies to establish a nationally coordinated climate modeling effort—instead of multiple climate models scattered across organizations—got no useful response. Several years later, after a successful World Modeling Summit in 2008 that I had helped organize, I tried again, still with no success. The US agencies engaged in climate modeling insisted and continue to insist, without a clear scientific justification and with considerable waste of national scientific and computational resources, that each agency needs its own climate model.

After the successful completion of the TOGA program, I wrote to the World Climate Research Programme—the governing body that organized and managed TOGA—to establish a global climate variability program that continues to this day. However, my proposal to launch a global climate experiment, just like Charney's Global Weather Experiment, one designed to study global ocean-atmosphere-land interactions and their predictability, never took off.

Prime Minister Modi didn't respond (or act on) the letters I sent urging him to persuade Donald Trump not to withdraw the United States from the Paris Climate Agreement, nor did he write back when I strongly suggested India launch a comprehensive national climate assessment, which is necessary for climate adaptation.

Not all my letters have been science-related. Once, in the run-up to the Iraq War, I wrote to both Jimmy Carter and Nelson Mandela and urged them to stage a hunger strike in protest—preferably in front of the statue of Mahatma Gandhi in downtown Washington, DC. I never heard back from President Carter but did get a response from Mandela's staff. "Unfortunately he would not be in a position to intervene in this matter," I was told.

Ten years ago, Anne and I were having dinner in a Virginia restaurant and got to talking with the waitress; she told us about a recent experience she'd had in which a group of foreign tourists had come to the restaurant, stayed for hours, and run up a large bill in the hundreds of dollars. Because tipping was not the custom in their country, they had not left her a gratuity. As she told us this story, tears filled her eyes. Even though she had worked very hard that night, she had made only $2.13 an hour, the minimum wage for servers. The conversation affected me so deeply that when we returned home, I immediately began looking for a nonprofit that fought on behalf of restaurant workers. I found One Fair Wage (led by the inimitable activist Saru Jayaraman) and emailed them immediately. In the ten years I have served on the organization's board of directors, it has changed the policies of several cities and is now campaigning for a national policy on tipped minimum wage.

And then there was the RICO letter, which didn't work in the way I had intended; the Justice Department did not take our suggestion, and *I* was the one subjected to an investigation. But I still count the letter as a success. My story, as much as it has been shared, as well as similar stories exposed by the Climate Science Legal Defense Fund became prime examples of the ruthless suppression apparatus at the heart of the climate-denial movement. Sometimes I tell my undergraduate class about the RICO saga, and I am always a bit flabbergasted when they applaud at the end. In the decade since I sat gloomily reading Congressman Whitehouse's op-ed, things *have* changed, and there is good reason for hope—starting with my students past and present, American and Indian.

Visiting my village these days is such a source of joy. Gandhi College has transformed the lives of thousands of rural students, especially young women. Parents of these women proudly declare to the families of prospective bridegrooms that their daughters are graduates of Gandhi College. Over the years, friends and fellow professors from GMU, Oxford, University of Washington, Seattle University, and Georgia Tech have traveled to Mirdha to visit the college and meet with the students.

As COLA rounds the corner on its thirtieth birthday, it's gratifying to see several of its scientists now professors and world-renowned leaders of their own research areas, and some of them are leading other scientific institutions. The Climate Dynamics Ph.D. program at GMU also continues to produce new experts in climate science—more than sixty students have received Ph.D.s so far.

And each year, the undergraduates who fill my Climate 101 course are more informed and more passionate than the students who came before them. In fact, during the ten years preceding Covid, the course has more than quadrupled in popularity.

These young adults have grown up beneath the looming shadow of catastrophic climate change, and let me tell you, they are not interested in fixating on the problems; they want to *fix* them. A few years ago, my students told me that the class was dwelling for too long on climate impacts, on fires and floods and habitat loss, and they requested we spend more time focusing on solutions. When I changed my last two lectures, which are now called "hopeful signs of climate solutions," they very much approved. It won't be long before these young people take the helm of companies, nonprofits, and governments, and I have faith that when they get there, they will make better decisions than their predecessors.

They are one of the reasons why I continue to teach, even as I mark my eightieth birthday—and one of the reasons I am much more optimistic about the future than most people would expect a climate scientist to be.

When I was growing up, the weather dictated my life: the food I ate (or didn't), the water I bathed with (or didn't), the days I spent hidden in a mango tree or curled in my mother's lap. As I got older, I moved between climate-controlled apartments and cities with snowplows, and the weather seemed less like an adversary to be feared and more like a curiosity, something to be understood, to be dissected and taxonomized. Eventually, when my academic interests turned to seasonal climate and beyond, I lost interest in the day-to-day weather altogether.

I'm not the only one. For many decades now, humans have lived as if *we* were in charge, finding ways to work around the weather. We've constructed levees, built large-scale irrigation systems, settled in places our ancestors could survive in for only seasons at a time. No one and nothing tells us what to do.

Except the weather is beginning to demand our attention. The brutal heat, the unprecedented floods: extreme events like these are no longer relegated to only the poor and far-flung populations of the world. The weather that arrives on my suburban doorstep every day is once again capricious and chaotic—just as it seemed when I was a village boy.

I know this causes many people to experience climate anxiety, and I do not blame them. I know firsthand how the world can feel like a cruel and random place. Things are unlikely to get better soon, and extreme weather is here to stay. But there is reason for hope.

I believe that to manage and mitigate climate change we need three things, and the good news is that we already have the first two well in hand. First, we need to understand the science. Check. Second, we need the technology that allows us to stop pumping the air full of carbon dioxide. Check. Third, we need *the will* to listen to the science and embrace the technology. It is only on this last point that we are stuck, thanks to the corporate greed that has parasitized our political system.

Just forty years ago, society found itself in a very similar situation. Chlorofluorocarbons, a harmful greenhouse gas used in foams, aerosols, and air conditioners, had torn a hole in the ozone layer, the planet's natural protection against the sun's damaging radiation. Fixing it required listening to the scientists issuing dire warnings, developing new technologies, and calling for action. In 1987, just two years after the hole was detected, forty-six countries entered into the Montreal Protocol, committing to phasing out harmful chlorofluorocarbons. In 2008, it was the first and only UN environmental agreement to be ratified by every country in the world. Today, virtually all ozone-depleting substances have been phased out of production, and the ozone hole is expected to close by the 2060s.[1]

Although it often feels like we are subject to a never-ending highlight reel of disasters and news that makes us feel hopeless, there really are, just like the story of the ozone hole, *hopeful signs of climate science* to be found in the world, reasons to stay optimistic. Sales of electric cars are booming. In the United States and Europe, emissions of CO_2 peaked in 2005 and 1979, respectively.[2] Wind farms are slowly rising over the Atlantic Ocean, and more than 80 percent of new utility-scale energy sources in America are renewables.[3] Recently, a group of teenagers sued the state of Montana for jeopardizing their right to a clean and healthful environment—and won! The road ahead is long, but we have already started our way down it.

I may be biased, but I would argue that meteorology represents our species' longest and most concerted effort to take care of one another. Indeed, it is easy to forget all we have accomplished. Five hundred years ago, humans couldn't begin to guess what caused the wind to blow or the sky to burst open with rain; it was the gods, they surmised, or the stars or the exhalations of the earth itself. One hundred years ago, the idea of producing an accurate weekly weather forecast was an impossible fantasy. Fifty years ago, the notion of predicting the climate months in the future wasn't just laughable—it was thought to be a scientific impossibility.

And yet, a tiny glimmer of hope inside a boy from a remote corner of India—nurtured by teacher after teacher—became the scientific basis for that very notion. Today, because of that boy's optimistic hunch, societies around the world are able to plan for the future, to prepare for and head off disaster. My own work was possible only because of the many indefatigable dreamers who came before me, the scientists who figured out the laws that govern the atmosphere, a method for finding order amid chaos.

It is true that the future scientists who are growing up today will face perhaps a more profound challenge than ever before. But as someone who has witnessed, time and time again, brilliant scientists gather from every corner of the globe and set aside political, religious, and cultural differences to overcome seemingly insurmountable challenges—I have faith they will rise to the occasion. Plus, tomorrow's meteorologists and climate scientists have some pretty neat technology on their side: machine learning and artificial intelligence, not to mention smarter satellites and faster computers, which will get even faster with quantum computing, enabling Earth system models of staggering resolution.

In the end, though, it won't be the models that save us. Nor will it be the supercomputers or the IPCC assessment reports or the climate NGOs. It will not be the climate scientists alone. It will be every person who takes personal responsibility for future generations and chooses to act, whether we do that by changing our consumption, volunteering our time for environmental nonprofits, casting our votes for politicians who prioritize climate action, or even just sitting down to write letters when we are so moved. As I tell my students, the best response to climate anxiety is climate action.

Our climate is changing, and our planet is in trouble. The path ahead can sometimes feel frightening and uncertain. But I have found predictability in the midst of chaos, a safe place to retreat when the storms are raging and the butterflies swarm. For me, it has been in the steady and unending work of making the world a better place.

I hope you will join me there.

Acknowledgments

The person I am today is thanks to the love and generosity of my mother, and the ambition, risk and resilience of my father.

I grew up in a remote, backward village as one of the six surviving children of my father, Chandrashekhar Shukla, and his second wife, Sita Devi. My village had no roads, no toilets, no school, and no electricity. Yet, after my father passed away, my mother and my elder brother Mahendra were determined that, despite the modest family resources, I should attend a college and a university.

I would like to thank my immediate family for their love and support—my wife Anastasia (Anne), my daughters Sonia and Pooja, and our granddaughters Natasha, Aarushi, and Aastha. (I would especially like to thank Anne and Sonia for their patience with my frequent discussions about the meaning of life.) I am grateful to my brothers Mahendra, Kanhaiya, and Shriram, and my sisters Vimla and Subhadra, as well as their families, for their love, affection and unwavering support. My brother Kanhaiya and his wife Manju have been a source of comfort for me during some of my most difficult days.

My friends Arthur and Susie Bass, Mark and Barbara Cane, Suraj Gyani, Santosh and Kiran Jiwrajka, Ken Mooney, Tim and Gill Palmer, David and Susan Straus, their daughter Emily, Mike and Susie Wallace, and Peter Webster have not only been a source of inspiration for my research and social work, but they have also traveled to my village to visit Gandhi College and inspire the students. I am especially grateful to the Jiwrajka family for their generosity of establishing a library at the college, and to Ramesh Modi, Naresh Modi, Ashok and Usha Padia, and Harsh Padia for creating a Home Science program at college. I am grateful to Susie Wallace for creating an English and computer learning program and teaching for several years, and to Emily Straus and Sonia Shukla who also taught at the college. Thanks to Vaija and Dileep Wagle for their sustained support and encouragement.

There is a long list of people to be acknowledged who have helped me in big and small ways during my professional journey from that backward village to Banaras Hindu University, to MIT, to NASA, to COLA, and to George Mason University. My apologies to anyone I have inadvertently left out.

It has been my good luck to learn from some of the great scientists in the field of weather and climate. By sheer luck, my four Ph.D. advisors were Jule Charney, Norm Phillips, Suki Manabe, and Ed Lorenz. It continues to be a privilege to have regular discussions with Suki Manabe, now a Nobel Laureate. My interactions with Abdus Salam and Roger Revelle inspired me to work with countries in the developing world.

I remain indebted to A. K. Tiwary and B. K. Tiwary, without whose support and encouragement I would have run away from BHU and back to my village. My BHU classmates from 1960, Kabir Roy Choudhury, Tarkeshwar Lal, R. N. Singh, and S. N. Thakur, have always encouraged me, and I enjoy our regular Zoom meetings to this day.

At Indian Institute of Tropical Meteorology, where my meteorological career began, I am grateful to my very first supervisor, K. R. Saha, and my colleagues D. A. Mooley, D. R. Sikka, and R. Suryanarayana, who accepted a boy from a village into their midst. I remain indebted to T. Nitta and K. Gambo from the Japan Meteorological Agency who, during my visit to Japan in 1967, introduced me to Taroh Matsuno and a two-layer model of the atmosphere. It was work with this model that sent me to the Tokyo conference in 1968 where I met Jule Charney.

I would like to thank the Fulbright fellowship program that made it possible for me to travel from India to MIT. The letter of admission to MIT from department chair Norm Phillips had made it very clear that I must have at least $500 with me when I enrolled. I am grateful to R. C. Srivastava, who lent me the money. His check arrived at MIT even before I did in August 1971. Werner and Judy Chasin volunteered to be a host family for a foreign student; I was that lucky student.

Thanks to my classmates, and the postdocs and professors in the meteorology department of MIT I never had a second thought about the wisdom of resigning a nice government job in India and coming to MIT as a graduate student. I would like to thank Arthur and Susie Bass, Mark and Barbara Cane, Dean Duffy, Inez Fung, Cathy Gebhard, Jerry Herman, Jeff Kroll, Antonio Moura, Ed Sarachik, John Walsh, and John Willett for their help and support in conquering my culture shock. Until this day some of us, including George and Hilda Philander and Dave and Tema Halpern, continue to meet, share meals, and laugh. It was at MIT that I met Ed Schneider and David Straus, two postdocs of Charney, who helped establish COLA.

A large contingent of Indian students at MIT enriched my life with shared meals followed by poker games and Bollywood movies. Veer Bhartiya, Anil Bhandari, Nanaji Saka, V. Krishnamurthy, Shobhana Rishi, Ashok and Usha Padia, Santosh and Kiran Jiwrajka, Nalini Srinivasan, Manjeet Singh Kalra, Deep Joshi, Avatar Singh, Kaplesh Kumar, and I remain friends and frequently meet even now. I am grateful to my roommate and my lifelong friend V. Krishnamurthy; we continue to collaborate on monsoon research. Anne and I regularly met Veer and Purnima and Anil and Dolly for meals until Veer passed away; now the five of us continue the tradition.

I am grateful to Norm Phillips, who arranged for me to spend one year at GFDL to work with Suki Manabe. I owe many thanks to GFDL scientists, especially Kirk Bryan, Bram Oort, Isodoro Orlanski, and Joe Smagorinsky, for their wisdom and guidance. During my postdoc days in Princeton, working with Doug Hahn showed me the importance of snow cover for seasonal prediction. I would also like to acknowledge the help and support of Bhopi Dhall who was in charge of the Texas Instruments supercomputer at GFDL, and thank Bhopi and Kamla for our lifelong friendship.

The 1968 meeting I had with Charney changed my career and my life. Charney helped me gain admission at MIT and taught me to be intellectually independent while I was there. I owe a huge debt of gratitude to Charney and Lorenz for their guidance and encouragement. I am also grateful to Henry Stommel, who planted the seeds of the idea that sea surface temperatures over the Arabian sea might influence monsoons.

It was also Charney who offered me a visiting faculty position to work at MIT and NASA, first at Goddard Institute for Space Studies in New York, and later at Goddard Space Flight Center, in Greenbelt, Maryland, where I had the good fortune to meet Jerry Herman and Michael Ghil, who also had similar appointments. I am grateful to Milt Halem, who offered me a senior federal job although I was not a US citizen. At Goddard, I was also the beneficiary of constant encouragement and support from Dave Atlas, Eugenia Kalnay, Bob Atlas, Joel Susskind, Yale Mintz, Piers Sellers, Dave Randall, Jerry North, Joanne Simpson, Bill Lau, and Yogesh Sud. It was here that research by Halem, Kalnay, Atlas, and Susskind to assimilate satellite data gave me the idea for reanalysis.

I would like to thank and express my gratitude to Jim Kinter, Brian Doty, Mike Fennessy, Larry Marx, Dan Paolino, Sumant Nigam, Ed Schneider and David Straus, who were my colleagues and collaborators at University of Maryland. I am most grateful to this group of scientists who helped me in advancing three major ideas: reanalysis, demonstrating the important role of land-atmosphere interactions, and establishing a scientific basis for dynamical seasonal prediction. Ed Schneider, Larry Marx, and Dan Paolino came to India to help me implement India's first global data assimilation and forecast system for medium-range weather forecasting. Drs. Moura and Krishnamurthy helped start the weather and climate training activity at ICTP in Trieste.

I am grateful to Jim Kinter, Ed Schneider, and David Straus for accomplishing what many thought was impossible: establishing an independent research center for seasonal prediction. An independent COLA would not have been possible without Jim Kinter and Anne Shukla. Jim Kinter helped create an environment of research at COLA that became the envy of the world; Anne Shukla created a financial management system that stood the test of the time. I was particularly encouraged by David Straus's decision to resign a secure federal job and join a soft money–funded COLA. After COLA left the university, Tim DelSole, Cristina Stan, Yongkang Xue, and several of our Maryland Ph.D. students—David DeWitt, Paul Dirmeyer, Bohua Huang and Ben Kirtman—joined COLA and transformed it into a world-renowned center of excellence for climate research. For me it is a daily joy to be surrounded by people much smarter than me.

In COLA's new home at George Mason University, it has been an honor working alongside COLA scientists who came to GMU, and Barry Klinger, V. Krishnamurthy, Paul Schopf, Emilia Jin, Natalie Burls, Kathy Pegion, Erik Swenson, and Debanjana Das, who joined the department. I remain grateful to Peter Stearns, Vikas Chandhoke, and later, Ali Andalibi, all of whom helped us transition to the university, creating the new Department of Atmospheric, Oceanic, and Earth Sciences, and establishing the Climate Dynamics Ph.D. program, a program that has graduated fifty-five doctoral students. I remain grateful to two brilliant students, Abigail Kokkinakis and Molly Reed, who were excellent teaching assistants.

Like most teachers, I have perhaps learned as much from my students as they have learned from me. For their help and support, I would like to acknowledge D. Achuthavarier, S. Amini, W. Anderson, K. Arsenault, A. Bamzai, S. Bates, D. Benson, M. Biasutti, R. Burgman,

I. Colfescu, D. Das, D. DeWitt, P. Dirmeyer, V. Dubey, L. Fudale, O. Gozdz, A. Hazra, H. Hsu, Y. Hou, B. Huang, B. Kirtman, E. Jin, Y. Jin, J. Joshi, E. Lajoie, J. Manganello, B. Narapusetty, C. Nobre, P. Nobre, V. Nolan, X. Pan, K. Pegion, P. Peng, O. Reale, R. Singh, B. Singh, A. Srivastava, E. Swenson, C. Tanajura, L. Xu, P. Yadav, R. Yang, J. Zhou.

Finally, I would like to thank Lauren Kurtz, executive director of the Climate Science Legal Defense fund, who introduced me to my agent, Lauren Sharp of Aevitas Creative Management. Lauren has expertly guided this process from proposal to finished project. Lauren also introduced me to my collaborator, Ashley Stimpson. This book could not have been written without Ashley, who I now consider a great friend. My idea of writing a book was to list all events sequentially as I do for my research. Instead, Ashley asked me to go back in time and describe my thoughts and feelings. She transformed my haphazard thoughts and recollections into beautifully written prose. I am deeply grateful to Lauren, Ashley, and the entire St. Martin's team, including editor Pete Wolverton, assistant editor Claire Cheek, and copyeditor Tracy Roe, who helped make this book a reality.

Notes

ONE

1. "The Global Atmospheric Research Programme (GARP)." OpenSky. Accessed December 1, 2022. https://opensky.ucar.edu/islandora/object/archives%3A9297.

2. Loveland, George A. "Definitive Short Range Weather Forecasting." *Bulletin of the American Meteorological Society* 8, no. 10 (October 1927): 153–156, JSTOR.

3. *Daily Weather Map Displayed at Smithsonian*. 1858. Smithsonian Institution Archives. Photograph. https://siarchives.si.edu/collections/siris_sic_514.

TWO

1. Nagaraj, Anuradha. "'Chaotic' Monsoons Threaten India's Farmers Without Climate Action." Reuters, April 14, 2021. www.reuters.com/article/us-india-monsoon-climate-change/chaotic-monsoons-threaten-indias-farmers-without-climate-action-idUSKBN2C117F.

2. Schilling, Mary Kaye. "India's Drought Is Killing Crops—And Pushing People to Suicide." *Newsweek,* August 16, 2018. www.newsweek.com

/2018/08/24/india-drought-suicides-climate-change-farmers-skulls-heat-disaster-1072699.html.

FOUR

1. Laskow, Sarah. "The Very First Forecast." *The Atlantic*, November 20, 2014. www.theatlantic.com/technology/archive/2014/11/the-very-first-forecast/382911/.

SIX

1. Norman A. Phillips. "The Emergence of Quasi-Geostrophic Theory." In *The Atmosphere—A Challenge*. Edited by Richard S. Lindzen, Edward N. Lorenz, and George W. Platzman, 177–206. Boston: American Meteorological Society, 1990. https://link.springer.com/chapter/10.1007/978–1–944970–35–2_11.

SEVEN

1. Witt, Stevens. "The Man Who Predicted Climate Change." *The New Yorker*, December 10, 2021. www.newyorker.com/news/persons-of-interest/the-man-who-predicted-climate-change.

ELEVEN

1. Sellers, Peter, Amnon Doucher, Yale Mintz, and Yogesh Sud. "A Simple Biosphere Model (SiB) for Use within General Circulation Models." *Journal of Atmospheric Sciences* 43, no. 6 (March 1986): 505–531.

2. NASA. "Taking a Global Perspective on Earth's Climate." https://climate.nasa.gov/nasa_science/history/.

3. Bengtsson, Lennart and Jagadish Shukla. "Integration of Space and In Situ Observations to Study Global Climate Change." *Bulletin of the American Meteorology Society* 69, no. 10 (October 1988): 1130–1143. https://doi.org/10.1175/1520–0477(1988)069<1130:IOSAIS>2.0.CO;2.

4. Woods Hole Oceanographic Institution. "1982–1983 El Niño: The Worst There Ever Was." www.whoi.edu/science/B/people/kamaral/1982–1983ElNino.html.

TWELVE

1. Shukla, Jagadish, Carlos Nobre, and Piers Sellers. "Amazon Deforestation and Climate Change." *Science* 247, no. 4948 (March 1990): 1322–1325. 10.1126/science.247.4948.1322.

THIRTEEN

1. Acharya, Nachiketa, and Elva Bennett. "Characteristic of the Regional Rainy Season Onset over Vietnam: Tailoring to Agricultural Application." *Atmosphere* 12, no. 2 (February 2021): 198. https://doi.org/10.3390/atmos12020198.

2. Tall, A., M. Simon, A. Maarten, P. Suarez, Y. Ait-Chellouche, A. Diallo, and L. Braman. "Using Seasonal Climate Forecasts to Guide Disaster Management: The Red Cross Experience during the 2008 West Africa Floods." *International Journal of Geophysics 2012*, no. 2 (January 2012). doi:10.1155/2012/986016.

3. NOAA. "Three-month Outlooks OFFICIAL forecasts." Last modified September 3, 2015. www.cpc.ncep.noaa.gov/products/predictions/long_range/seasonal.php?lead=1.

4 NOAA. "US Seasonal Drought Outlook." Last modified March 22, 2024. www.drought.gov/data-maps-tools/us-seasonal-drought-outlook.

FIFTEEN

1. Witze, Alexandra. "Our Climate Change Crisis." *Science News,* March 10, 2022. www.sciencenews.org/century/climate-change-carbon-dioxide-greenhouse-gas-emissions-global-warming#the-first-climate-scientists.

2. Revelle, Roger, and Hans E. Suess. "Carbon Dioxide Exchange Between Atmosphere and Ocean and the Question of an Increase of Atmospheric CO_2 during the Past Decades." *Tellus* 9, no. 1 (February 1957): 18–27. https://doi.org/10.1111/j.2153-3490.1957.tb01849.x.

3. Weart, Spencer. "Climate Change Impacts: the Growth of Understanding." *Physics Today* 68, no. 9 (September 2015): 46–52. https://doi.org/10.1063/PT.3.2914.

4. Elliott, W., and L. Matcha. 1979. "Carbon dioxide effects: research and assessment program." Presentation at the workshop on the global effects of

carbon dioxide from fossil fuels, Miami, Florida, March 7, 1977. www.osti.gov/servlets/purl/6385084.

5. National Research Council. *Carbon Dioxide and Climate: A Scientific Assessment*. 1979. Washington, DC: The National Academies Press. https://doi.org/10.17226/12181.

6. Elliott, W., and L. Matcha. 1979. "Carbon dioxide effects: research and assessment program." Presentation at the workshop on the global effects of carbon dioxide from fossil fuels, Miami, Florida, March 7, 1977. https://www.osti.gov/servlets/purl/638508.

7. Huddleston, Amara. "Happy 200th Birthday to Eunice Foote, Hidden Climate Science Pioneer." Climate.gov, July 17, 2019. www.climate.gov/news-features/features/happy-200th-birthday-eunice-foote-hidden-climate-science-pioneer.

8. Shepherd, Marshall. "How a Woman You Never Heard of Helped Enable Modern Weather Prediction." *Forbes*, January 24, 2017. www.forbes.com/sites/marshallshepherd/2017/01/24/how-a-woman-you-never-heard-of-helped-enable-modern-weather-prediction/?sh=6f7dd893195c.

9. Sur, Abha. "The Life and Times of a Pioneer." *The Hindu*, October 14, 2021. https://web.archive.org/web/20140413141835/http://hindu.com/2001/10/14/stories/1314078b.htm.

SIXTEEN

1. IPCC. "Structure of the IPCC." Accessed March 23, 2024. www.ipcc.ch/about/structure/.

2. Randall, D. A., R. A. Wood, S. Bony, R. Colman, T. Fichefet, J. Fyfe, V. Kattsov, A. Pitman, J. Shukla, J. Srinivasan, R. J. Stouffer, A. Sumi, and K. E. Taylor. "Climate Models and Their Evaluation." In Solomon, S., D. Qin, M. Manning, Z. Chen, M. Marquis, K. B. Averyt, M. Tignor, and H. L. Miller (eds.) *Climate Change 2007: The Physical Science Basis. Contribution of Working Group I to the Fourth Assessment Report of the Intergovernmental Panel on Climate Change*. Cambridge, UK, and New York: Cambridge University Press.

3. Gallant, Allie, and Sophie Lewis. "Lost in Translation: Confidence and Certainty in Climate Science." *The Conversation*, August 22, 2013.

https://theconversation.com/lost-in-translation-confidence-and-certainty-in-climate-science-17181.

4. IPCC. *Climate Change 2007: Synthesis Report. Contribution of Working Groups I, II and III to the Fourth Assessment Report of the Intergovernmental Panel on Climate Change*. 2008. Geneva, Switzerland: Intergovernmental Panel on Climate Change. www.ipcc.ch/site/assets/uploads/2018/02/ar4_syr_full_report.pdf.

5. Merrill, Abigail, and Grace E. Hirzel, Matthew J. Murphy, Roslyn G. Imrie, and Erica L. Westerman. "Engaging the Community in Pollinator Research: The Effect of Wing Pattern and Weather on Butterfly Behavior." *Integrative and Comparative Biology* 61, no. 3 (September 2021): 1039–1054, https://doi.org/10.1093/icb/icab153.

6. van Hasselt, Sjoerd, Roelof Hut, Giancarlo Allocca, Alexei L. Vyssotski, Theunis Piersma, Niels C. Rattenborg, Peter Meerlo. "Cloud Cover Amplifies the Sleep-Suppressing Effect of Artificial Light at Night in Geese." *Environmental Pollution* 573 (January 2021): published online, doi:10.1016/j.envpol.2021.116444.

7. Omand, Melissa M., Deborah K. Steinberg, Karen Stamieszkin. "Cloud Shadows Drive Vertical Migrations of Deep-Dwelling Marine Life." *Proceedings of the National Academy of Sciences of the United States of America* 118, no. 32 (August 2021). https://doi.org/10.1073/pnas.2022977118.

SEVENTEEN

1. Brulle, Robert. "Institutionalizing Delay: Foundation Funding and the Creation of U.S. Climate Change Counter-movement Organizations." *Climatic Change* 122 (February 2014): 681–694. https://doi.org/10.1007/s10584-013-1018–7.

2. Walker, Joe. "Draft Global Climate Science Communications Plan." Memorandum to API's Global Climate Science Team. www.documentcloud.org/documents/784572-api-global-climate-science-communications-plan.html. Accessed March 26, 2024.

3. Leiserowitz, A., E. Maibach, C. Roser-Renouf, G. Feinberg, and S. Rosenthal. "Climate Change in the American Mind." New Haven, CT: Yale Proj-

ect on Climate Change Communication. 2017. https://climatecommunication.yale.edu/publications/global-warming-ccam-march-2015/.

4. Bastasch, Michael. "Scientists Ask Obama to Prosecute Global Warming Skeptics," *Daily Caller,* September 17, 2017. https://dailycaller.com/2015/09/17/scientists-ask-obama-to-prosecute-global-warming-skeptics/#ixzz4VDSPViaL.

5. P. Greg. "Alarmists Enraged: 20 Climate Scientists Ask Obama to Use the RICO Act to Go After Global Warming Skeptics." *twitchy*, September 17, 2015. https://twitchy.com/gregp/2015/09/17/alarmists-enraged-20-climate-scientists-ask-obama-to-use-the-rico-act-to-go-after-global-warming-skeptics-n458080.

6. Briggs, William M. "Failed Climate Scientists Call For RICO Investigation To Stop Criticisms, And Non-Scientist Claims Scientists Will Cause Next Genocide." *William M. Briggs: Statistician to the Stars!* (blog), September 18, 2015. http://www.wmbriggs.com/post/16865/.

7. Delingpole, James. "Climate Scientist Caught in 'Largest Scandal in US History.'" Breitbart, October 2, 2015. www.breitbart.com/politics/2015/10/02/climate-alarmist-caught-largest-science-scandal-u-s-history/.

8. Smith, Lamar. Letter to the author, October 1, 2017. https://science.house.gov/_cache/files/4/7/474438c2–0842–4d3a-8120–98b5c94b2fa9/2B0B259715A9322FBF9F72F7D7D7DA7C.10–1–15-cls-to-shukla.pdf.

9. Union of Concerned Scientists. "Appendix B: Groups and Individuals Associated with ExxonMobil's Disinformation Campaign." *Smoke, Mirrors & Hot Air: How ExxonMobil Uses Big Tobacco's Tactics to Manufacture Uncertainty on Climate Science*. Union of Concerned Scientists, 2007. JSTOR. http://www.jstor.org/stable/resrep00046.10.

10. Corriher, Billy. "Trump's Anti-Environment Judicial Nominees Could Lead to Polluted Air and Water." *American Progress*, July 27, 2017. www.americanprogress.org/article/trumps-anti-environment-judicial-nominees-lead-polluted-air-water/.

11. "Lamar Smith," Open Secrets. Accessed March 27, 2024. www.opensecrets.org/members-of-congress/summary?cid=N00001811&cycle=CAREER.

12. Abraham, John. "Lamar Smith, Climate Scientist Witch Hunter." *The Guardian*, November 11, 2015. www.theguardian.com/environment/climate-consensus-97-per-cent/2015/nov/11/lamar-smith-climate-scientist-witch-hunter.

13. Gillis, Justin. "Clouds' Effect on Climate Change Is Last Bastion for Dissenters." *The New York Times,* April 20, 2012. www.nytimes.com/2012/05/01/science/earth/clouds-effect-on-climate-change-is-last-bastion-for-dissenters.html#:~:text=The%20scientific%20majority%20believes%20that,climate%20researcher%20at%20Texas%20A%26M.

14. Goldenberg, Suzanne. "American Tradition Institute's fight against 'environmental junk science.'" *The Guardian*, May 9, 2012. www.theguardian.com/environment/2012/may/09/climate-change-american-tradition-institute.

15. Schouten, Cory. "Who Files the Most FOIA requests? It's Not Who You Think." *Columbia Journalism Review*, March 17, 2017. www.cjr.org/analysis/foia-report-media-journalists-business-mapper.php.

16. "Sunday Reflection: The Collusion of the Climate Crowd." *Washington Examiner*, July 7, 2012. www.washingtonexaminer.com/sunday-reflection-the-collusion-of-the-climate-crowd.

17. Halpern, Michael. *Freedom to Bully: How Laws Intended to Free Information Are Used to Harass Researchers.* Cambridge, MA: Union of Concerned Scientists, 2015. www.ucsusa.org/resources/freedom-bully#.VN4xeC7ooWU.

18. Burakoff, Maddie. "Confidence in Science Fell in 2022 While Political Divides Persisted, Poll Shows." Associated Press, June 15, 2023. https://apnews.com/article/trust-science-medicine-social-survey-725ab3401f27900be6cc957eec52e45e.

EIGHTEEN

1. United Nations. "Rebuilding the Ozone Layer: How the World Came Together for the Ultimate Repair Job." Last modified September 15, 2021. www.unep.org/news-and-stories/story/rebuilding-ozone-layer-how-world-came-together-ultimate-repair-job.

2. Shendruk, Amanda. "Tired of Feeling Hopeless About Climate Change? Take a Look at These Graphs." *The Washington Post,* September 6,

2023. www.washingtonpost.com/opinions/2023/09/06/climate-change-charts-data-optimism/.

3. Fasching, Elesia. "Wind, Solar, and Batteries Increasingly Account for More New U.S. Power Capacity Additions." *Today in Energy, U.S. Energy Information Administration*, March 6, 2023. www.eia.gov/todayinenergy/detail.php?id=55719.

Index

Terms of relationship such as "Wife" are in relation to the author, Dr. Jagadish Shukla.

About the Author

Tom Wakefield

Dr. Jagadish Shukla is a Distinguished University Professor of Climate Dynamics at George Mason University. Internationally recognized for his role in the advancement of weather and climate science, he has received the Rossby Medal, the Walker Gold Medal, and the International Meteorological Prize by the UN, and the Medal for Exceptional Scientific Achievement of NASA, the highest honor given to a civilian by that agency. He was one of the lead authors for the Intergovernmental Panel on Climate Change's 4th assessment, for which he and his team shared the Nobel Peace Prize with Vice President Gore.